Knowledge Management Case Book

Siemens Best Practises

Edited by Thomas H. Davenport and Gilbert J. B. Probst

Second Edition, 2002

Publicis Corporate Publishing   John Wiley & Sons

Die Deutsche Bibliothek – CIP-Cataloguing-in-Publication-Data
A catalogue record for this publication is available from Die Deutsche Bibliothek

 INKNOWVATE! is the registered trademark for
Siemens Corporate Information and Operations
Knowledge Management (CIO KM) activities.

As a combination of the verbs innovate and know
it represents the final stage of the CIO KM roadmap
(initiate, mobilize, institutionalize, innovate).

http://www.publicis-erlangen.de/books
http://www.wiley.com

ISBN 3-89578-181-9

Second Edition, 2002

A joint publication of Publicis Corporate Publishing and John Wiley & Sons

Editor: Siemens Aktiengesellschaft, Berlin and Munich
© 2002 by Publicis KommunikationsAgentur GWA, Erlangen

Printed in Germany

Knowledge as a competitive advantage

Foreword by Dr. Heinrich v. Pierer, CEO of Siemens AG

"Knowledge is no good if you don't apply it", said Goethe. And he could have been talking about us. For, today, companies like Siemens have to exploit their expertise more systematically and more intensively than ever before. Between 60 and 80 percent of the value added we generate is linked directly to knowledge – and the proportion is growing. As a result, one of our company's first priorities is to network and manage our internal knowledge so that we will be even more efficient and provide even greater benefits to our customers.

There aren't too many problems that one or the other of our business segments hasn't already solved. Whether it's installing a complete metropolitan subway system, constructing a pharmaceuticals plant on a turnkey basis or putting up an office tower with the latest building management and communications technology – you can bet that at least one of the 450,000 Siemens experts in at least one of the 190 countries where we are active has tackled the job before.

This treasure trove of experience is one of our key competitive advantages. But we have discovered that not enough of our knowledge is being fully exploited. And so we have devised a number of measures to improve our knowledge utilization.

A team at company headquarters pools information about the knowledge management activities of our entire company, develops new methods for implementing knowledge management, defines further goals and coordinates projects, thus creating a company-wide network for knowledge exchange and organization.

The technical basis for this effort is provided by our worldwide company intranet. For example, our Information and Communication Networks Group has established a ShareNet linking sales, marketing and service personnel around the world twenty four hours a day and enabling over 20,000 information and communications technology specialists in more than 80 countries to provide one another with immediate assistance and customers with instant access to Siemens' global know-how. We have decided to build upon this success to create a Siemens-wide system, linking all company employees worldwide.

But, by itself, excellent technology is not enough. Above all, our employees have to be open to one another, ready and willing to learn and share their knowledge. We have, therefore, taken a number of concrete steps to encourage cooperation. For example, we have, as part of our *top*⁺ program, reorganized the way we share information, experience and know-how. Through these measures, we are on the road to becoming a real learning company: an organization that absorbs ideas from the outside world by regularly benchmarking its activities against those of its best competitors and that shares best practice examples internally, holding up world-class achievements for everyone to emulate. Our ultimate goal is to ensure that all of our people can access the company's enormous pool of knowledge. And this is what knowledge management is all about: it is people business. That means the experience and abilities of our people are – and will continue to be – of ever-greater importance for our company's competitiveness and profitability.

Our success in the area of knowledge management has been duly recognized in the annual assessment made by the heads of the Fortune Global 500 companies and knowledge management experts around the world. Siemens is the only German corporation to have been ranked among the world's top 20 knowledge management companies in each of the past four years. In 2001, we were Number 1 in Germany, Number 2 in Europe and among the top ten worldwide.

This second edition of the Knowledge Management Case Book once again provides concrete examples of the way Siemens is fostering, promoting and optimizing knowledge utilization. The individual case studies included are also a valuable source of ideas for efficient, targeted knowledge management.

Munich, January 2002

Dr. Heinrich von Pierer
President and CEO of
Siemens Aktiengesellschaft

Contents

7

IV Added Value of Knowledge Management

V Learning and Knowledge Management

VI Visualizing more of the Value Creation

VII Epilogue

Written by

Introduction

Siemens' Knowledge Journey

Thomas H. Davenport and Gilbert J. B. Probst

The movement by Siemens AG toward Knowledge Management over the last two or three years, has been – at least to external observers – a quite extraordinary transformation. This book illustrates, by way of case studies, how this major multinational transformed itself into a knowledge-based company. The case studies in this book, written by Siemens managers with the assistance of young researchers, provide a stimulating and captivating account of a traditional company getting ready for the new economy.

With this book we invite you to join us in exploring Siemens' transformation into a knowledge-based company – its knowledge journey. Here we have one of the world's oldest, largest, and historically most successful corporations. Yet, here we also have a firm that is able to transform itself from within with a minimum of commotion. In the eEconomy, today's core competencies become tomorrow's core rigidities with unprecedented speed. Under these conditions it is incumbent upon companies to ruthlessly reconsider the value of established processes and ways of doing business, and to gear the organization towards purposeful attendance of its most valuable asset – its knowledge base. In a quiet and pervasive way, Siemens has become an organization that is highly dependent upon knowledge to advance its strategies and key business processes. A firm, once noted for its bureaucracy and its deliberation, has wholeheartedly embraced a new business concept within a brief period of time.

In this foreword, we will describe why the management of knowledge is particularly suited to this venerable company and why we do believe the Siemens transformation is unique. We will conclude with some challenges that Siemens may face in its future pursuit of Knowledge Management. An overview of the structure of the book is also provided to guide you in your reading of the individual cases.

Why Knowledge Management makes sense at Siemens

There are many good reasons why Knowledge Management has found a receptive home within Siemens. First of all, the firm is truly global, which means that for employees to share knowledge they must do so through means other than (or in addition to) informal face-to-face communications. The aspects of Knowledge Management that involve technology-enabled repositories and sharing networks – that is, the parts that help to overcome geography – were well suited to an organisation as dispersed around the world as Siemens.

Siemens is, of course, a company built on technology. Increasingly, information technology is either a part of, or an important facilitator of, Siemens' diverse businesses. Since Knowledge Management is greatly enhanced by the effective use of information technology, it's not surprising that Siemens was a relatively early and enthusiastic adopter of Knowledge Management. The IT-driven nature of the company's businesses also provides a strong motivation to manage knowledge effectively. One attribute of these technologies is that they change very rapidly; keeping up with various computing and communications technologies is much easier when a company has a system for rapidly circulating new knowledge. But Knowledge Management at Siemens goes far beyond purely IT-based systems, as you will soon discover in this book.

Siemens is also a highly diverse organization that participates in a wide variety of businesses. In its history, the company has been called a conglomerate and for decades business and organization scholars have deliberated on how such collections of relatively independent businesses could generate synergies, or increased value through collaboration. How can the whole be made greater than the sum of the parts? Firms hesitate to ask individual business units to help one another for fear that they will suboptimize their own performance. However, Knowledge Management offers a potential solution to this dilemma. If knowledge can be easily shared across business units, then one Siemens business unit can take advantage of the learning and expertise of another.

Of course, this sharing requires more than just technology and a common communications network. There must be a willingness to share, and a sense that employees are all part of one broad organization, not just their own business units. These factors are settling firmly into place at Siemens. Several of the firm's knowledge-oriented "Communities of Practice" (for examples of this see Part II. of this book) cut across business units, and the "Best Practices Marketplace", maintained by the Corporate Knowledge Management group, elicits participation from employees in multiple units. In these ways, Siemens is beginning to enjoy a "knowledge synergy" across a highly divisionalized organizational structure. The cases in this book describe how this synergy is developed and maintained.

But there is another important rationale for moving toward Knowledge Management: the general change experienced in customer offerings, not just at Siemens, but also at many other manufacturing-oriented firms. Instead of simply offering highly capable manufactured goods, Siemens very early on understood that most of its products are knowledge-based and therefore demand appropriate and systematic ways of caring for this knowledge. In the end, the customer is the primary beneficiary of this process of purveying a variety of value-added services. Products themselves become part of "total solutions", including services that meet the customers' needs. Many of these services are knowledge-based, and benefit from activities that capture, distribute, and apply knowledge. All these activities are covered in this book.

Over time, we expect that many Siemens business units will become more knowledge-based. Knowledge will become embedded in the product and service strategies of all the units. Significant portions of the value Siemens provides to its customers will be a result of knowledge, both tacit and explicit. This transformation has already begun, and will continue to characterize Siemens' direction for the next several decades. The cases in this book elucidate the way and the milestones of getting there. In many ways, therefore, the present book serves as a roadmap of the knowledge journey of this company.

The uniqueness of Siemens' approach

In the past, Siemens was known for its strong hierarchy, but its approach to Knowledge Management has not been hierarchical at all. Instead the approach has been from the grassroots up and therefore relatively "bottom-up". Without suggestion or prompting from above, a number of mid-level employees and managers of Siemens business units began to create repositories, Communities of Practice, and informal sharing approaches for knowledge. In most business units the opinion of those taking the initiative was simply that the time had come to begin managing knowledge. The fact that Knowledge Management was "the right thing to do" was sufficient justification for them to make a start.

After communication and many quick wins, the employees looked around and noticed that others were doing the same thing. After a period of informal communication, the employees and managers who were managing knowledge began to form a semi-official Community of Practice themselves. Clear messages from the top management convinced them that they were on the right track and that there was a need for a coordinated approach of a Siemens-wide management of knowledge. As Heinrich von Pierer observed in the early 1990s, "50 percent of the value-added at Siemens comes from knowledge-intensive products and services". In the preface to this book he now emphasizes that today between 60 and 80 percent of the added value generated is linked directly to knowledge – and the proportion is still growing. The Corporate Knowledge Management function, established in recognition of the need for a coordinated approach to Knowledge Management, still thrives on the community concept from which it emerged. At Siemens therefore, most of the knowledge management efforts still take place in the business units, but the corporate group plays a valuable coordinating role.

We recently observed, for example, that more than 200 people attended a firm-wide Siemens conference and workshop of knowledge managers from various business units. We have not seen such a large group of knowledge managers convene within any organization other than consulting firms. The knowledge sharing and exchange that took place at this gathering was an impressive display of bottom-up coordination.

Siemens' approach to Knowledge Management is also unusual for the diversity of initiatives and applications that are underway within the company. Most firms that we have observed, focus almost all of their efforts on one major initiative – most commonly a knowledge repository. At Siemens however, the variety of initiatives and applications – many of which are described in this book – is much greater. There is also a wide variety of knowledge content domains that are addressed within the firm, including "best practices", customer knowledge, competitive intelligence, product knowledge, financial knowledge, and so forth. The scope of approaches and tools being employed across Siemens corresponds to the diversity and complexity of the organization itself.

Siemens' approach to Knowledge Management is also notable for the extent of its relationships with external entities, particularly universities. This book, for example, was produced by Siemens in collaboration with doctoral students from different universities (the Ludwig Maximilian University of Munich, the Universities of Graz, St. Gallen, HEC-University of Geneva and MIT). At Siemens case writing on Knowledge Management was therefore not only a natural event, but also a knowledge management tool in itself. In the last chapter, Gilbert J. B. Probst will comment on this approach, the participants' role in group writing, as well as the effectiveness of this as an organizational learning tool. Projects members and coaches (mostly internal, but in collaboration with professors in the knowledge management field) realized that best practices should be shared within the company's network. Siemens understands that in the current, networked, eEconomy, companies that harvest and hoard their knowledge will be at a competitive disadvantage. Companies today live in knowledge ecologies where one company feeds knowledge into another. What counts is a networked approach to Knowledge Management, involving internal as well as external parties. The logic behind this is as simple as it is compelling: if you cut off the outflow of knowledge, you will also cut off the inflow. We believe, therefore, that the firm's openness to external experts and the sharing of ideas within a broad network will be a key driver in maintaining competitive success at Siemens.

How Siemens mastered the challenges of Knowledge Management

Making the transition to becoming a firm that manages all aspects of knowledge well, is clearly going to be difficult. In some ways, the firm's ability thus far to have mastered these challenges is evidence of its commitment and resolve.

For example, the importance of knowledge and Knowledge Management to Siemens is best illustrated with the following: Siemens executives have begun to guarantee shareholders that they will achieve a better-than-market rate of return from capital invested in the firm. The company also plans to be listed on equity markets in the United States – a set of capital markets requiring high levels of profitability, growth, and financial return. This is, therefore, a relatively difficult time for Siemens to be making substantial investments in intangible capabilities such as Knowledge Management, but the company has found a way to do this. Even at a time when investments are becoming subject to a higher level of scrutiny, Siemens business units and the Corporate Knowledge Management function have found the resources to get Knowledge Management off to a good start.

Another challenge is charting a course between local and global knowledge initiatives. While earlier we argued that the global nature of Siemens provides a strong rationale for Knowledge Management, it also presents difficulties. As you will observe in this book, Siemens business units have knowledge initiatives that cut across countries and continents as well as other programs that work only within a particular country. Employees must choose to allocate their attention and energies to either a global system, or a local one. Many participants may appreciate the value of a global system, but may also be more comfortable with the language of their own country. The important balance between local and global Knowledge Management thus needs to be maintained on a daily basis. A prime example of successful management thereof is the company's knowledge-sharing network, "ShareNet". Case number 1 describes how ShareNet achieves a balance between global networking and local relevance.

Another tension in companies aspiring to become truly knowledge-based is between knowledge initiatives that support the entire firm, versus those that advance a particular business unit, or an even smaller group within it. Firm-wide initiatives help to exploit the scale of Siemens' business and promise the "knowledge synergy" we described above. More specialized, focused initiatives are easier to measure and may be better supported by managers who are responsible for a unit's financial performance. Like the local vs. global distinction mentioned above, this is a creative tension that will play out over time. Thus far, Siemens managers have handled it – better than many of the large, multi-business firms we have encountered.

A final challenge to knowledge management at Siemens is how to nourish it during difficult economic times. When knowledge management first emerged at the company, most of the world economy was in growth mode, as was Siemens. After the new millennium, Siemens (and its competitors) experienced substantial decreases in demand for some of its product segments, particularly those involving information technology and telecommunications. Yet knowledge management has continued to survive and even to thrive. It has been recognized as a tool that can facilitate efficient and productive work processes, as well as innovation and growth. In a sense, then, the second edition of this book is a testimony to the power and persistence of knowledge as a critical Siemens resource.

Suggestions for reading this book

The cases in this volume are the result of a collective effort of Siemens employees, and (external) coaches – we called them "case coaches" – who acted as devil's advocates in the process of conceptualizing and writing up the cases. Specific emphasis was given to the isolation of key propositions towards the end of each case study. These provide, in a condensed form, the gist of the entire case. Thus, if you are a manager, interested mainly in the implications of a case study for your company and your approach to managing knowledge, the key propositions are a good place to start reading the cases. Together with the abstract at the beginning of each case study, the key propositions will give the hurried reader a good overview of the case study at hand. The case studies have all been written in a narrative style. We found that this not only enhances the accessibility of the account, but also benefits the transmission of the knowledge. The narrative style is more authentically conducive to portraying the rich experiences and knowledge in the cases than a neutral, "academic" style. We trust you will agree.

Each case was written as a complete and independent study. Internal trainers, consultants and professors wishing to use the Case Book for instruction, can select individual cases without losing the richness of the entire book. Hence, the cases in this book can be read or used in any order. This also means that some redundancy is unavoidable, and indeed intended. Each case provides the context necessary for fully appreciating its content and message, regardless of previous cases. The reason individual cases were not presented as consecutive chapters, with one building on the other, was to reinforce the idea that they do not form a linear pattern, and can hence be read according to your preferences.

The book is organized in seven parts. The book starts by making a fundamental connection that provides a conceptual frame for the entire book: **knowledge and strategy**.

The *Knowledge Strategy Process (KSP) case* introduces the KSP as a strategic instrument for the business owner and his/her management team, which should be integrated into the business strategy process and revisited regularly. The resulting KM action plan is a guideline for the KM team and a very valuable contribution to the company's KM roadmap that is strengthened by the buy-in of the management team through the KSP. A knowledge strategy for a business is defined by the business owner and his management team in six steps. These steps lead from the currently most relevant business perspective for the near future, related key performance indicators and knowledge areas to assessing the "knowledge area state" (as-is and to-be) in proficiency, diffusion and codification, based on a comprehensive understanding of "knowledge" in the business. Finally, KM actions are defined to achieve to-be states for prioritized knowledge areas yielding state-of-the-art KM solutions, with the latter being focused by business objectives and orchestrated across all knowledge-related management disciplines.

Knowledge Management in the current business environment, where there is too much information, much of it inane, is very much about identifying the aspects of knowledge that are actually relevant. At the same time, we are all familiar with the idea that companies often know more than they can tell. Thus, the challenge revolves around firstly identifying relevant knowledge and then transferring this knowledge timeously to the parties that actually need it. Under the current business realities, striking the balance between identifying relevant knowledge and transferring it becomes the prerequisite for survival. Hence the present book places explicit emphasis on methods that can help to achieve this balance. Hence Part II contains four cases addressing fundamental issues in **knowledge transfer**.

The *ShareNet case* describes leveraging knowledge on a global basis as a major challenge to big multinationals like Siemens. Induced by significant changes within the international telecommunication business, Siemens Information & Communication Networks (ICN) anticipated a shift in competitive pressures that stressed the necessity for knowledge-based competition. Therefore the ShareNet case describes how Siemens ICN succeeded in its transformation from mainly a product seller into that of a global solution provider. It outlines the role ShareNet, a business application system, played within this transformation and discusses the critical success factors involved.

The *IndustrialServices case* looks at the sharing of knowledge specifically within a sales and service environment. In this environment, business is mostly driven by employees at the customer interface. By virtue of their intensive contact with the customer, sales representatives and service technicians are not only one of the major sources of knowledge and experience in these organizations, but also of future customer needs. After a short introduction to the organizational context, the market shift that created a need for the systematic management of knowledge is discussed. The study then gives an overview of a knowledge-management initiative – its approach, relevant factors, and some of the underlying design principles.

The *KN Service Knowledge case* describes the Know-how Exchange, a unique knowledge-sharing network tool connecting the employees of Siemens Industrial Services. An extensive database, it was compiled by those at the coalface of industry and pools their knowledge gained from practical experiences of diverse projects in more than three hundred locations throughout the world. This knowledge then becomes channeled for use by others via this innovative tool. The Know-how Exchange makes it possible for this invaluable knowledge to be shared and experiences to be reused, thereby increasing customer benefit and process performance. Expert knowledge, the fruit of many years' innovative work, is now transparent and essentially available via each employee's individual PC. The immediacy and practical application of this tool are related via an authentic narrative, interspersed with additional motivational and background information of its development, problems and current state.

The *Best Practice Marketplace case* investigates the establishment of a best practice sharing marketplace, i.e. a forum for exchanging best practices on corporate level. Specific attention is paid to the barriers to best practice transfer. Strategies for overcoming these barriers are illustrated and critical success factors for best practice transfer are also considered. The major finding of the case was that best practice transfer approaches have to be interwoven with employees' day-to-day work in order to be successful. The case concludes with a summary of the key propositions arising from the case and a list of questions for further reflection on the topic.

Communities of Practice are an important ingredient in the sharing of knowledge. Recent research points to Communities of Practice as the primary agents in enabling knowledge sharing. Several prerequisites need to be met in order to make such Communities successful. Perhaps the most critical among these is a balance of explicit management of Communities on the one hand, and of the deliberate self-organization of Communities on the other. How this balance can be achieved is the focus of Part III, and the Power of Communities, Infineon and CoP Support cases describe how Siemens addresses these issues.

In the *Power of Communities case* the phenomenon of Communities of Practice is examined as a driving force for effective Knowledge Management in a company. The challenges facing the set-up and successful implementation of a Siemens knowledge management community are explored. As a forum for sharing knowledge and a knowledge community, and as a testing laboratory for integral knowledge-management systems, its strengths and critical aspects are discussed. Thereafter, the factors that contributed to its successful creation are examined. In particular, the question of how this Community of Practice in Siemens – a non-centrally organized and heterogeneous company – was able to develop and attract sufficient attention to bring into being a new central corporate office, is examined together with its aim – to co-ordinate and support the Knowledge Management activities of Siemens.

The *Infineon case* describes how a manager and his team implemented and systematized the exchange of process knowledge within the factories of Infineon Technologies (manufacturers of silicon microchips), in five countries on three different continents, in order to solve manufacturing problems and to optimize procedures. The ambitious business goals of the company could only be realized if all ten sites were able to achieve the same level and quality of output. This would only be possible, if valuable expertise were easily, continuously and freely accessible, as and when the problems arose, and not only sporadically, as had been the pattern in the past. To this end, the Knowledge Exchange Networking (KEC) project was initiated in 1990, in cooperation with the Knowledge Management Department of Siemens Central Technology Division. The KEC, as the project became known, is ideally suited to serve the ever-changing and highly competitive field of semiconductor technology, where strategic technological knowledge has a relatively short "half-life" and thus needs to be transferred an-dimplemented as quickly as possible.

The *CoP Support case* describes how supporting Knowledge Communities is an ideal way of promoting knowledge creation and sharing across widely dispersed groups, regions and corporate centers within Siemens, as well as ensuring that Siemens is able to access the knowledge of the entire company at the right time. To this end the case first outlines and discusses important aspect of Knowledge Communities, such as the interactive process that forms the community lifecycle, important groups of role players, means of communication, and the components of successful Knowledge Communities. The history of the Siemens Knowledge Community Support is thereafter presented from the time Corporate Knowledge Management (CKM) started calling on experts from Corporate Technology, Siemens Business Services and Siemens Qualification and Training to pool their know-how and be partners in this endeavour. The development and realization of the comprehensive and detailed support of the Knowledge Communities within Siemens are described. The case closes with the current status and an overview of the newest developments.

Knowledge Management in itself does not necessarily produce superior value. But when knowledge is applied to marketable products and services, the true **added value of Knowledge Management** emerges. Such application can take different forms, and Part IV illustrates innovative arenas where such knowledge application can take place.

The *knowledgemotion case* illustrates the knowledge management framework that was designed for the introduction of knowledgemotion, the company-wide knowledge management initiative at Siemens Business Services (SBS). The knowledge management framework will give the reader an understanding of the integrated approach to

Knowledge Management and the different stages of implementation. It also introduces the key learning processes experienced by SBS during the various implementation phases of knowledgemotion. The knowledge management requirements, challenges and solutions within the service business are highlighted. The case study also shows the requirements, objectives, challenges and solutions of knowledge management programs, in general, and at SBS, in particular. Based on the knowledge management implementation experience at SBS, the case study closes with critical success factors for knowledge management implementations, both within and outside Siemens.

The *MED case* demonstrates Siemens Medical Solutions' (MED's) knowledge management approach, overall concept and implementation of knowledge management solutions along the value chain, and the integration of these solutions into the processes and the daily workflow. Each solution is described in detail and the potential value thereof is defined for every target group. The increasingly systematic management of knowledge in a proactive division such as Siemens Medical Solutions is explained. MED will continue to be motivated by its end goals, namely to secure short-term success and long-term viability. A particular knowledge management objective in the support of whichever strategy the division pursues, is to leverage the best available knowledge to make people, and therefore Siemens Medical Solution itself, as effective as possible. An integral part of this case study is illustrating how MED achieves its goal of dealing with operational, customer, supplier, and all other challenges while implementing its strategy in practice.

The *Mergers and Acquisitions case* investigates a knowledge management approach to mergers and acquisitions and looks as the unique challenges this poses for a company. The establishment of a knowledge exchange to meet these challenges led the Mergers & Acquisitions Knowledge Exchange (MAKE) team at Siemens to develop a comprehensive and dynamic knowledge base that is linked to real and virtual access to expertise in this increasingly important and growing aspect of corporate life. Insights gleaned from this experience will provide food for thought for others who find themselves caught up in, but ill-equipped to deal with, the multi-faceted subject of mergers and acquisitions.

For Knowledge Management to be applied thoroughly, prior and ongoing learning is required. Part V illustrates the various areas where **learning and Knowledge Management** are taking place within the company.

The *Knowledge Master case* illustrates the need for and benefits of a corporate learning program for Knowledge Management, as well as its implementation. The realization of the program, called Knowledge Master, through a public-private partnership is described, as are the program design and virtual-learning platform. Participants' experience of the program is also reviewed. The study closes with a look at critical issues in the design of successful corporate learning programs on Knowledge Management.

The *Management Learning case* comprises a general survey of the background, approach and structure of the Siemens Management Learning programs. Founded in 1997 these programs marked a radical new beginning in management education. Today they have become crucial for knowledge management within the company and represent the most important force for worldwide strategic management development within Siemens. The case study, moreover, provides several examples illustrating the procedure and results of the Business Impact Projects that form – beside workshops and e-learning – a major part of the programs. The build-up and purpose of the projects are first explained, where after concrete examples as well as business results and key learnings are provided. Finally several key propositions enhance the transfer possibilities to other business contexts.

The *E-learning case* is based on the fact that a growing number of companies and organizations, realizing that learning faster than the competition is the only competitive advantage which is viable, are developing a new learning culture. However, getting an organization's learning capacity to gain momentum is not that easy.. How do you do this and how do you sell your approach to an entire company? This case study offers some thoughts on the essence of learning organizations and how e-learning and Knowledge Management can help them to become a learning organization. It places the 'Siemens Learning Valley' for Belgium and Luxembourg in the right perspective. It's a story about the birth, the vision, the strategy, the results, the challenges and the future of Siemens Learning Valley as a project that drives the learning power of the organization.

Knowledge Management has pointed to the increasing importance not only of transferring knowledge and applying knowledge, but also to the importance of measuring the contribution of core knowledge and its management to corporate success. In this perspective, it is critical to evaluate the added value that Knowledge Management provides in some quantifiable manner. You can only manage what you can measure. After all, **visualizing more of the value creation** is a critical basis for developing incentives for further stimulating knowledge sharing and networking on local and global levels. Without quantifiability, measurement endeavors remains elusive. Further, it is critical to ensure that existing knowledge assets are constantly challenged in a purposeful way. Especially in the current "Internet Age", where today's core competencies quickly turn into tomorrow's core rigidities, it is incumbent upon companies to ensure that the knowledge they nurture inside is still relevant to the market. This is why Part VI provides three cases that explicitly address the issue of developing metrics and incentives for Knowledge Management.

The *Premium-on-Top case* describes how knowledge networking on a global level can be given a nudge. For this purpose, two projects were combined. On the one hand there was the Knowledge Network Model (LITMUS) project, through which international Knowledge Management is measured, and on the other hand there was the Bonus-on-Top, which makes worldwide knowledge transfer and creation attractive by offering employees valuable incentives. Thus case number 11 goes beyond mere measurement of knowledge networking, and suggests ways in which the results of such measuring can be used to promote further knowledge networking.

The *Knowledge Networking case* answers the question of how Knowledge Management can help transform an organization. This question became relevant when the German market demand for Siemens' information and telecommunication products changed dramatically for the worse, and its sales unit required a solution to this crisis required. This solution entailed a change from selling products to selling solutions. Top management decided to employ knowledge management to support this change. The newly founded Knowledge Networking department started creating an infrastructure of knowledge management tools, thereafter measuring their local usage as well as the frequency of individual networking. The Knowledge Networking department finally used the benchmark figures to initiate a local consultancy. During this last phase it became obvious that related themes had to be brought in to create a methodology of focused analysis for change and control phases, thus answering the initial question.

The *E-business transformation case* illustrates the challenges that will accompany new business in an increasingly dynamic and global company environment such as Siemens and explores some of the fundamental directions that empower the e-community. This case study offers a short introduction to and an overview of the scope and elements of the e-business transformation within Siemens. Knowledge Management, which is one objective within this e-business program, is the topic and focus of this case. It considers the motivation and objectives of the e-community as well as the applied processes and methods in e-business projects. It likewise discusses relevant e-business Knowledge Communities throughout the world and Sharenet, the workspace for e-excellence. Aspects of learning and the motivation of employees are described as well as the effects of cultural change. Completing the case are some examples of results achieved with Knowledge Management for the described e-business transformation. This deliberately "open ended" case study closes with a reflection on the key propositions that arose from this project and areas for future improvement. This case study also illustrates the challenges and objectives of knowledge management programs that are relevant for Siemens.

An *epilogue* by Gilbert J. B. Probst constitutes Part VII and concludes the book. This epilogue elucidates the methodology used in writing the cases. But it is more than just a methodological reflection. It affords a different layer of meaning to the cases in that it portrays the entire book as a knowledge management tool in itself. Hence, you will find that the individual cases provide accounts of how knowledge can be tended to successfully, and the book, itself, is yet another such tool. No wonder case studies have been used for some time now in MBA classes. But this book goes beyond mere instruction. The idea in the epilogue is to demonstrate that the very process of case writing is instrumental in managing knowledge within and among organizations. In the current, networked, business environment, such case writing can therefore serve as a convenient tool for creating and exchanging knowledge within your company and also with external partners.

As the cases in this book amply suggest, the Siemens approach to Knowledge Management is certainly no fad. There is a pervasive belief that Siemens is in the knowledge business, that Siemens workers are knowledge workers, and that an important part of management is Knowledge Management. Were a similar volume to be created a decade hence, we believe we would find that Knowledge Management has become so pervasive within the firm that it would be part of everyone's job, and difficult to isolate as an activity in itself. Perhaps the best future indicator of Siemens' success with Knowledge Management will be its invisibility!

Acknowledgements

At this point we would like to thank the individual contributors who made this book possible. Above all, our sincere gratitude to heads of Corporate Information and Operations at Siemens Dr. Fritz Fröschl and Chittur Ramakrishnan as well as to Siemens Knowledge Officer Günther Klementz, for their enthusiasm and encouragement in putting this book together.

Michael Gibbert, doctoral student at Siemens and Peter Heinold, Program Manager CIO KM Marketing&Communication have contributed greatly to the organization, the team-writing processes, the rationale and the ideas expressed in the first and second edition of this book. Without them, the books would not have been completed. The doctoral students Michael Franz, Uwe Trillitzsch, and Sven Völpel have to be mentioned for their enthusiastic cooperation. Special thanks are due to the following two persons: Gerhard Seitfudem from Publicis Erlangen, as our publisher his support, infinite patience and imperturbable personality are greatly appreciated, and even more so when deadlines are looming. Special thanks to Ilse Evertse for her linguistic and stylistic acumen and for "turning ducklings into swans" in the process of copy-editing. She provided the firm hand and inspiration needed to keep the contributors going whenever the project seemed to be flagging. In both editions of the book, she was much more than a copy editor. Furthermore many thanks to all the case coaches, and the Siemens employees who invested their time and energy in making this a successful and memorable event.

I Knowledge Strategy

The Knowledge Strategy Process – an instrument for business owners

Josef Hofer-Alfeis & Rob van der Spek

Abstract

The Knowledge Strategy Process (KSP) is a strategic instrument for the business owner and his/her management team, which should be integrated into the business strategy process and revisited regularly. The resulting KM action plan is a guideline for the KM team and a very valuable contribution to the company's KM roadmap that is strengthened by the buy-in of the management team through the KSP.

A knowledge strategy for a business is defined by the business owner and his management team in six steps. These steps lead from the currently most relevant business perspective for the near future, related key performance indicators and knowledge areas to assessing the "knowledge area state" (as-is and to-be) in proficiency, diffusion and codification, based on a comprehensive understanding of "knowledge" in the business.

Finally, KM actions are defined to achieve to-be states for prioritized knowledge areas yielding state-of-the-art KM solutions, with the latter being focused by business objectives and orchestrated across all knowledge-related management disciplines.

The method as well as experiences from the KSP's applications in Siemens AG is described in detail. The procedure in a KSP project, lessons learned, integration with measurements and further developments support the application of the method in other organizations.

Introduction

Siemens AG Corporate Knowledge Management introduced the Knowledge Strategy Process (KSP) into the corporation as a method for business owners and their teams to determine strategy and action plans. The KSP was tested as a pilot project by Siemens AG from January to October 2001 and found to be very useful. Its integration into existing strategic instruments, such as Balanced Scorecards, and into measuring techniques for knowledge and Knowledge Management (KM) is well underway, and the company-wide rollout is being prepared.

The following findings report begins by describing the reasons why a knowledge strategy is an indispensable prerequisite for implementing business-oriented, well-focused and cross-functional knowledge management. It then proceeds to outline the methods and experiences from the pilot project, illustrating both processes and findings. The Knowledge Strategy Process can be used as a direct and structured way to gauge how successful knowledge management is. This is discussed below, followed by a description of how the KSP is likely to develop and concluded by a discussion of lessons learnt and success factors.

The basic principles of the Knowledge Strategy Process (KSP) method were developed by the Dutch knowledge management company CIBIT in Utrecht [1] and licensed by Siemens. The companies joined forces to refine these basics [2, 3] and are currently working together on their further development. Furthermore the basic method has been applied in various other medium-size and large companies by CIBIT.

Knowledge strategy versus knowledge management strategy (KM roadmap)

The Knowledge Strategy is a dedicated instrument used by business owners and their management teams to plan, implement and control management actions concerning business-relevant knowledge. The latter, both as a resource and as a product, is having a growing impact on business success. The Knowledge Strategy identifies which knowledge areas have an impact on the business, how strong this impact is, which deficits there are in each of the knowledge areas in terms of proficiency, codification and diffusion, and determines what the management feels it can do in response to these issues.

A knowledge management strategy or roadmap is targeted at knowledge management managers (Chief Knowledge Officers or similar staff functions) and their cross-business responsibilities to enable KM. The KM roadmap describes the implementation, operation and standardization of basic components of KM solutions and change initiatives. This ensures that knowledge management systems can be efficiently and swiftly introduced into organizations and thereafter effectively operated. (Our experience is that KM systems are always socio-technological systems with one or more knowledge communities as the driving force of the sharing and creating of knowledge, which are the core processes of knowledge management.)

Actions that solve common issues in several knowledge areas, or across different businesses are a major output of a business-specific Knowledge Strategy. These actions, in turn, are a valuable requirement input to the KM roadmap of the Chief Knowledge Officer (CKO). Nevertheless, actions that concern only business-specific knowledge issues have to be implemented by the related managers in this business.

The KM roadmap may also contain actions which, from the CKOs perspective, are considered essential to enable KM solutions in the future, but which are not yet required by a knowledge strategy.

The difference between a knowledge strategy and a knowledge management roadmap also becomes clear if one recalls similar pairs of planning instruments for other important business factors, such as products, customers or quality. For example, top managers naturally have a product strategy, but they expect a product management strategy from the relevant product managers.

Why have a knowledge strategy and how is it formed?

Many KM projects have been and are bottom-up initiatives. Generally speaking, specialists or middle management, human resources managers, I&C managers or process managers are convinced that they will reap benefits if their unit has a knowledge management solution, which they usually create themselves or together with consultants. Experience has shown, however, that whenever considerable resources and investments for KM, or business and organizational transformations become necessary, or whenever the business situation becomes tougher, business owners usually prove unwilling to truly support the effort. Studies of knowledge management obstacles or success factors demonstrate this again and again. [4, 5]. In most cases, the greatest obstacles disappear and crucial success factors are guaranteed if the management (the business owner and his team) is able to design, align and monitor knowledge-related management activities according to its strategic perspective and business experience. In other words, when the business strategy incorporates the essential features of a knowledge strategy, success is guaranteed. This is precisely what a knowledge strategy process (KSP) does: it guides people in order to define the relationship between business development, key business indicators and the necessary knowledge areas (for knowledge portfolio, see the pilot example). It furthermore determines the actual and target statuses of these knowledge areas (for Knowledge Status Guide, see the pilot example). Based on their understanding of the business, the management and specialists then use the KSP to draw up an action plan for knowledge-related management actions in their own language. The resulting projects are then drafted and implemented – with the cooperation of an interdisciplinary knowledge management team and a KM consultant, if required. Thus orchestrated and state-of-the-art KM solutions can be achieved. All this serves to secure the management's engagement and leads to a situation where success can be directly monitored by reviewing the various specified actions, objectives and impact relationships in the knowledge strategy.

The above means, firstly, that KM projects will be oriented towards business objectives and that KM resources will be concentrated wherever they will have the greatest impact within the company. For example, people will use the KSP to determine where a knowledge community should be strategically supported, or where knowledge management instruments should be implemented to have the greatest business impact. Secondly, the potential beneficial synergies could be exploited and organizational inefficiencies avoided if KM and the other major knowledge-related management disciplines – e.g., human resources and continued education, I&C technology, organization and process management, but also strategy, research and innovation – cooperate to create KM solutions and solve problems.

These benefits are essentially due to the comprehensive knowledge model upon which knowledge management and the KSP at Siemens are based [for basics see 6] and which embodies the guiding principle: "all business-relevant knowledge is distributed".

The basic comprehensive knowledge model

The key dimensions of knowledge, which can be used to identify the actual or target status and any potential action needed, are:

- *Proficiency (abilities, skills and expertise)*, which are always tied to particular people
- *Diffusion*, which reflects to what degree abilities and expertise are distributed and how the processes for distribution and networking are working
- *Codification*, which conveys to what extent and how knowledge is documented or expressed in some other way.

For example, consider the multifaceted knowledge required to run a fine French restaurant. Although a great deal depends upon the chef and his or her outstanding expertise, the final result is also determined by the knowledge to be found in his or her team of cooks and the networks of the suppliers and customers. This does not even include the many recipes in published or in-house cookbooks, the processes or learning techniques.

If we take a closer look at the work methods and needs of today's knowledge worker, we see that here, too, it is vital that the distributed knowledge necessary to meet his or her objectives is properly managed. To a certain extent it is immaterial whether the knowledge worker obtains results as a result of his or her own capabilities, or from networking with peers or other specialists, from calling on the assistance of consultants, or by resorting to documented models, recipes and proceedings.

According to this definition, knowledge can be described as the capability for effective action that is to be found either concentrated in individuals, or distributed throughout organizations, and which goes hand-in-hand with information about knowledge, i.e. about that capability. Knowledge management therefore comprises all actions by means of which knowledge may better contribute to the success of the business.

This knowledge model and the examples also reveal how unproductive discussions are for successful knowledge management when they attempt to define and separate "knowledge" (as we understand it in our knowledge management) and "information", as well as other heuristic entities and opportunities.

The method and practical experience and pilots

Within Siemens AG, KSP was tested using practical applications in two pilot projects and several training sessions. To date the widest usage of KSP has been in Siemens ATEA, a subsidiary company of Siemens Belgium and Luxembourg. KSP has been applied in ATEA's research, development and engineering organization that employs approximately 700 developers in information and communications technology. The process and some results of this pilot project are described in more detail below, however, since the output of all these activities always includes confidential information, the results shown have been simplified and neutralized.

The basic KSP steps and the procedure during the pilot project

The Knowledge Strategy Process consists of six basic steps, which result in a knowledge management action and project plan. Figure 1 is a representation of the six steps.

In discussions with the business owner and his strategic team, the question posed in the first step of KSP is answered:

Step 1: What is the most significant business perspective for the near future?

This can be a new product line, a process innovation, a business or an organizational transformation. What is the timeframe planned for it? These considerations determine the direction the knowledge strategy will take. In the pilot, the strategy was oriented towards business as a whole and the pending expansion of the service business over the course of two or three years. Then the business owner's management team along with the department heads and staff came in. They had to be convinced of the wisdom of applying KSP. Only thereafter could the KSP workshop take place at an executive level. During this workshop the 12 participants and two KSP consultants made known the findings mentioned below.

Step 2: Which knowledge areas are significant for the selected business perspective?

For this process, a knowledge area constitutes the thematic consolidation of experiences, theories, findings and abilities in the various manifestations of the knowledge model described. It touches on considerations that can range, for example, from whether a company has the know-how to develop energy-saving engines right through to its available expertise to conclude projects successfully. The question is answered by the team during the process of brainstorming, structuring and selecting, and may take several hours the first time the KSP is performed. The result is a list of 10-12 knowledge areas the team has agreed upon. The results from the pilot are illustrated in part in the knowledge portfolio; see Figure 2 below.

Step 3 (often alternated with Step 2): Which of the key performance indicators used for business apply to the selected perspective?

This is where it becomes clear whether a business strategy has been explicitly formulated, because these indicators can simply be taken from that strategy, and without it, there is no point in having a knowledge strategy. In the pilot, for example, the indicators included customer success, a performance index for project execution, employee satisfaction and an innovation index.

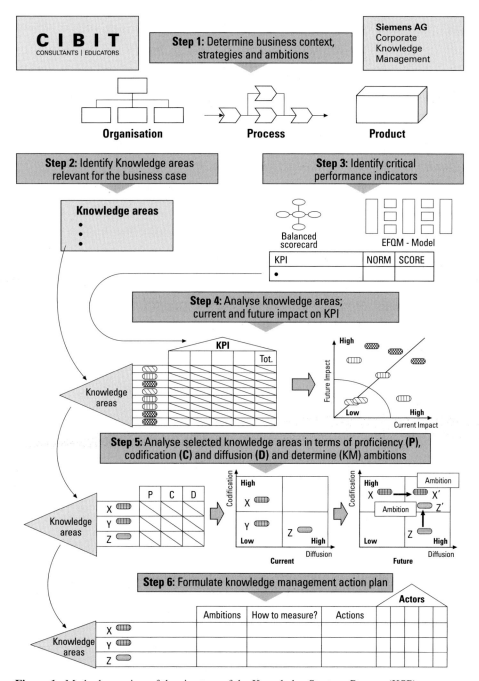

Figure 1 Method overview of the six steps of the Knowledge Strategy Process (KSP)

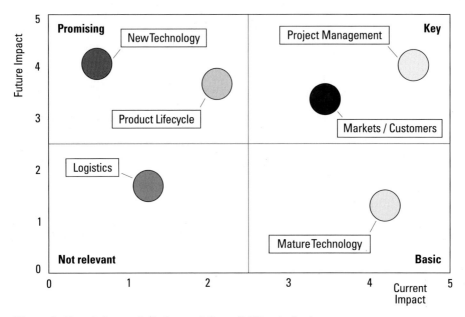

Figure 2 Knowledge portfolio (excerpt) for an R&D organization

Step 4: What is the current and future impact of the knowledge areas on the Key Performance Indicators?

This is where the team relates the results from the previous steps to one another by discussing and assessing the current and future impact of a knowledge area on a performance indicator. Knowledge areas are therefore subjected to a two-dimensional weighting according to the degree of business impact, which can be displayed in the knowledge portfolio.

The sample portfolio clearly shows that "project management", followed by "market or customers", are the knowledge areas with the greatest current and future impact on business. "Technologies" would, of course, also have been a key knowledge area with the pilot project's partner, but because it contains too many important individual knowledge areas, the team subdivided this area into two separate categories. They are: "mature technologies", located in the "basic knowledge areas" portfolio sector, and "new technologies" in the "promising knowledge areas" sector.

Unless the team already has some experience with developing a knowledge area list as a result of similar discussions about core competencies (which is often the case), the questions in Step 5 of the KSP usually require some getting used to. They delve more deeply into the comprehensive knowledge model, described above, and thus naturally require more time for explanations and discussion the first time KSP is used.

Step 5: What's the status of our knowledge areas and where should we improve?

This activity will focus on the "fitness" of the knowledge areas in terms of the three knowledge dimensions, namely proficiency, diffusion and codification. This is especially relevant to those knowledge areas considered most important for the selected business perspective, i.e. those in the knowledge portfolio which lie further to the right or higher up on the graph.

To ascertain this "fitness", the team estimates the actual and target status. In some cases this could require queries or initial research. Typical questions here are:

- Do we have experts, or at least one world-class expert, working independently in this field of knowledge? (And who are they?) Competence management procedures can be used to conduct more detailed investigations, if necessary.

- Is proficiency distributed among all the relevant stakeholders who need it, and at what level? In the pilot project, the relevant stakeholders entailed, for example, various expert laboratories, but also included functions at every point along the entire business process, such as manufacturing or sales, as well as the essential external knowledge carriers, e.g. suppliers or customers. The purpose of this question is to determine how knowledge is distributed and thus how well the processes associated with knowledge diffusion are working.

- What is the status of our codification? Are there: Reports? Structured descriptions? Standardized, coordinated models, such as best practices?

A status chart called Knowledge Status Guide is derived from the answers to these questions and illustrated in Figure 3.

The Guide shows the actual and target status of a knowledge area, both in terms of diffusion and codification, with the head and tail end of a corresponding arrow line. The

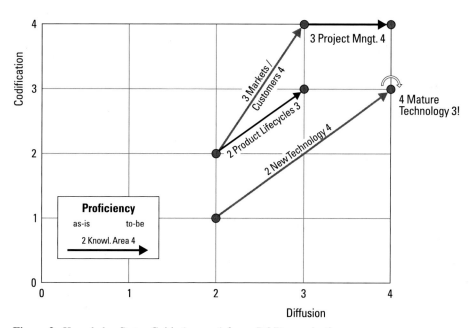

Figure 3   Knowledge Status Guide (excerpt) for an R&D organization

third dimension, the status of proficiency, is represented by the number before and after the knowledge area label next to the line. The Guide therefore expresses the team's overall estimate of the need for change and, together with the knowledge portfolio, the two act as an orientation for developing proposals for knowledge management actions. In this step, as in the previous steps of the KSP, the decision-making process is just as critical as the findings in the process of encouraging insight and intent to change in the team. For this reason, the discussions during the process are a key component of KSP documentation. This is where managers' support for KM projects is gained (an element often lacking), since they are given the option of developing the causal links between business objectives and KM actions.

Step 6: What's our plan and how do we monitor our progress?

This is where conclusions are drawn from the analysis as to what is to be done with knowledge. It will also become immediately clear that, from a business perspective, isolated initiatives by employees in the various knowledge-based disciplines, such as managers in HR, learning and training, and information and communication as well as organization and process designers have to be integrated. During this process, knowledge and innovation managers should support this collaborative effort with their integrative concepts to achieve orchestrated KM solutions. This procedure corresponds precisely to the comprehensive model of distributed knowledge, which always has the potential to open more powerful "multi-dimensional" creative opportunities.

Step 6 can be divided into three sub-steps between which prioritization steps can be inserted wherever necessary.

Step 6a: Suggestions for actions and a detailed supportive analysis

During its "business talk", the management team develops proposals for actions from its strategic perspective and experience with knowledge-related actions. During this process it may be necessary for very large, important knowledge areas to run additional knowledge strategy processes. This has to be done by additional teams, consisting of management and specialists, in separate workshops, according to the relevant detailed level. During the pilot, two more KSP workshops were held for each of the knowledge areas in technology. During these workshops some 30 detailed knowledge areas were analyzed and actions proposed for them. Naturally, additional subject matter experts participated in these detailed workshops.

Step 6b: Integration of actions

After the detailed workshops, the actions proposed therein are prepared by KSP consultants and grouped into cross-knowledge area and area-specific actions. Thereafter these actions are rated in the overall KSP context and presented to the management team. In the pilot, the following examples were selected from approximately 100 individual suggestions:

- Six basic actions, which were given top priority and high-level management attention.

- Approximately 15 cross-knowledge area actions. This was an important outcome for the KM officer, or the interdisciplinary KM team and the cross-business KM road-map.
- Specific action packages, which were adopted for all the knowledge areas analyzed. These actions would be the responsibility of either the managers for that knowledge area, the related knowledge community and its moderator, or a corresponding organizational function, if in place.

The following were the most interesting results from the pilot in this regard:

- The introduction of a new knowledge area-related discipline, namely that of the "Domain Manager" who then coordinates and integrates KM actions from the various disciplines.
- The development of standard terminology for the entire business to prevent this from being continuously recreated for various detailed actions, e.g. for the mapping and organizing of knowledge communities or specialists, as well as for various content management activities.
- A revision of the strategy process: Of particular significance to the management was the finding that the knowledge portfolios from the detailed workshops, which resulted from the increased participation by specialists, sometimes differed significantly from the business strategy plans. The result was that corresponding strategy planning actions had to be organizationally more effectively integrated with one another.

Other examples of actions proposed:

- Providing time for sharing knowledge and learning
- Establishing the model of the time-sharing expert who gradually migrates from one mature knowledge area to a new one (for the "mature technologies" knowledge area)
- Establishing better learning processes between R&D and sales
- Starting a company academy for the systematic transfer of key knowledge
- Providing an overview of regional and corporate knowledge communities and improving the communication between these communities
- Creating a general methodical framework for both the "solution development" and "development integration" knowledge areas
- Constructing a marketplace for "lessons learned" (structured findings reports).

Step 6c: Design and planning of solution strategies

The actions and resolutions proposed from the business perspective are translated into KM approaches on the basis of findings, or analyses of the actual status of KM systems or planning. This basically means that KM projects for state-of-the-art solutions are planned and implemented in all the disciplines involved. This is where the "bottom-up" KM introduction process and the "top-down" KSP approach intersect, but with the great advantage that the management knows what it wants and why. KM officers in the pilots occasionally had their reservations, fearing that already existing planning and projects would be "scrapped" as a result of the inclusion of the management team with its privileged insights into knowledge issues of the business. In retrospect, however, people were generally enthusiastic, because the clear objectives and the changed intentions of the executive level energized and accelerated KM actions more than ever before. In addition, a Chief Knowledge Officer would normally not have the authority to enforce changes such as, for example, those concerning fundamental organizational or business transformations.

More key results from the pilot project

In addition to the planning results discussed in the above paragraph, the following impact generated by the KSP pilot in Siemens ATEA is also of interest:

- The KSP will be implemented in marketing and sales and other organizational units in Siemens ATEA and will be regularly conducted together with the business strategy process.
- The person previously responsible for knowledge and innovation management in the R&D organization was made responsible for KM and became the KSP process owner for the whole company.
- The regional companies' headquarters is currently considering introducing the KSP.
- The integration of the KSP into the very advanced processes of the Balanced Scorecards in this region will be jointly investigated.

Configuration of the KSP

The Knowledge Strategy Process can be performed using a set of interviews and workshops. The number of workshops will depend on the organizational level on which one starts. A common approach could be:

- An interview with the business owner to identify the major issues in the context of the business case (step 1). During this interview consensus should be achieved about the boundaries of the business case, which people should be involved and the preparation that could be initiated for the activities that will follow.
- An initial workshop bringing together relevant stakeholders on a business level. This workshop should be focused on the specific six steps, but issues should be raised on a general level. This workshop should result in a clear understanding of the key performance indicators and knowledge areas that are important on a business level, but should also give direction to further analysis. An outcome of this workshop could be a selection of specific knowledge areas that require further steps. Another outcome could be that the analysis focuses on a different or additional perspective (i.e. a specific process or product).
- Steps 3 to 6 can be repeated for each appointed knowledge area by involving different stakeholders. This will lead to more detailed results and a better focused action plan. Our experience is that this in-depth analysis requires two or three workshops, since the level of detail requires more analysis activities and information gathering.

- In the case of a different or additional perspective, steps 1 to 6 should be performed again, but this time involving different stakeholders.

Though we have performed several successful one-day workshops at which we executed all 6 steps, experiences demonstrate that dividing the process into several activities (partial workshops) achieves optimal results. This provides the moderators, the leadership team and staff involved with the opportunity to prepare the various steps, to digest the outcomes, to eventually investigate specific issues or underpin educated guesses by means of analysis and data gathering.

Lessons learnt from and success factors for knowledge strategies and the KSP approach

Based upon the pilots within Siemens AG and the experiences of CIBIT within other companies, the following lessons learnt were identified:

1. The KSP will always be an iterative process. During the process stakeholders will share their perspectives, information will be gathered and analyzed and the result will hopefully be a shared vision. This process cannot be linear, but should allow iterations based on ongoing insights.
2. The added value of this approach lies, first of all, in the process in which various stakeholders share their opinions and perspectives. Communication between these stakeholders is as important as the outcome. It is therefore important to involve all relevant stakeholders.
3. It is possible to start on a general level (i.e. company level) and to repeat the process for more specific knowledge areas or specific processes, units or products. However one should keep in mind that results on a general level will also be "general" and refer to clusters of knowledge (i.e. knowledge about customers, markets, products, project management, HRM etc.). One should always consider whether this is necessary in order to focus the KSP on specific knowledge areas.
4. This process should be integrated into the strategic management cycle and repeated when necessary. The business environment will change and the knowledge management actions should respond to this change as soon as possible.
5. It is crucial to use existing and well-established key performance indicators in this process. It is advisable to relate this process as far as possible to the corporate performance measurement system.

Based upon these lessons learnt several critical success factors were identified:

- *The business owner must be convinced*

  There is the often extremely difficult task of convincing business owners that they will have vast new opportunities if they are willing to expand their business strategy by adding a knowledge strategy, although this may require considerable effort. The initial issues regarding KSP are clarified in preliminary talks with business owners and their planning team.

- *The KSP workshop team should be correctly composed*

  The right mixture of management, topic specialists, and knowledge-related staff functions is crucial for the enforceability and quality of the strategic decisions.

- *The most important disciplines of knowledge-related management activities must be included*

  The KM officer should integrate the other players into an interdisciplinary knowledge team for synergetic integration of the individual initiatives to occur.

- *The differences and similarities between knowledge strategy and the KM roadmap must be clarified*

  When the KSP is used on a broad scale, a knowledge strategy is used for each business in the business units, but only one KM roadmap is required for all the businesses. The cross-knowledge-area KM actions from the knowledge strategy are often actions that several businesses need and thus should be elements of the KM roadmap. On the other hand, KM action planning in the KM roadmap can itself specify suggestions and the state of the art that can be built on.

- *Professional process consulting and documentation*

  The KSP is a process with issues which are still new to people. In order to moderate intense discussions efficiently, a KSP advisor must have team moderation skills (especially for management teams), experience with using KSP, and a wealth of ideas. Identifying decision-making processes requires another aid: an Excel-based tool, ideally run by a KSP assistant, was developed for documenting results in the KSP steps. The essentials of the (often heated) discussions can be documented best if handled by an employee with knowledge of the subject and business particulars.

Knowledge and knowledge management metrics and the refinement of the KSP

The question of impact

Metrics for knowledge and knowledge management, as they relate to KSP, will only be discussed briefly here. The knowledge manager usually hears the following remarks from the business owner: "You can certainly implement the knowledge management project, but I expect to see results from very early on." The KSP changes that situation completely: The benefits of the KM process are pinpointed during the KSP, and the management follows this by drawing up a plan of action. At the same time, the causal links in the KSP analyses act as metrics of KM impact on business objectives, essentially working in the reverse direction.

- KM actions and projects are measured against corresponding sub-objectives and success indicators. For example, has an overview of available knowledge communities been made and communicated? To what extent have users deployed it and with what degree of success?

- KM actions are focused on reducing deficits in knowledge status indicators (see the Knowledge Status Guide), accordingly improvements must be visible the next time the KSP is performed.
- Since the influence of the knowledge areas on key performance indicators (KPIs) is defined in the KSP, the effect of KM actions in the knowledge area can also be examined for their impact on the indicators. In such a case, it must of course be clear that because numerous factors, e.g. further change initiatives, are intertwined with key performance indicators, only a rough assessment is possible.

These connections also demonstrate potential links between the KSP and Balanced Scorecards that are at present being investigated.

Irrespective of this, other metrics are available to the Chief Knowledge Officer (CKO) and the KM team as important instruments for:

- Measuring the KM system performance
- Summarizing the development status of the business transformation into a world-class company, e.g. by means of corresponding benchmarking or a "degree of KM maturity" survey.

These will not be discussed.

Potential developments

There are several ways in which the KSP can be refined:

- Methods for teams to identify and structure knowledge areas. We will build upon various techniques that have developed in disciplines, such as knowledge engineering, creativity and content management.
- Further elaboration of metrics for the status of knowledge areas, instruments and KM-projects.
- Expansion of the Balanced Scorecard, or business excellence models, by adding knowledge areas, knowledge statuses and business impact indicators. This expansion should ensure that the output of the KSP is directly linked to the performance measurement system.
- Elaboration of additional dimensions in the knowledge status framework:
  - Creation of new knowledge (innovation): prerequisites and metrics,
  - Identifying new business opportunities based on knowledge, or knowledge management systems which exist in the knowledge areas,
  - Identifying knowledge risks, i.e. business risks that have not been dealt with previously in connection with the knowledge areas.
- KM action templates that fit certain knowledge status constellations and business features. We have learned that actions are closely related to the lifecycle of knowl-

edge areas with regard to their business impact. Emerging knowledge areas require actions that differ from those that mature knowledge areas require. This relationship will be explored in more detail in the future.

References

[1] Van der Spek, Rob, Kingma, Jan: "Achieving Successful Knowledge Management Initiatives," in: *Liberating Knowledge*, business guide of the Confederation of British Industry, published by Caspian Publishing, 1999
Contacts: Rob van der Spek & Jan Kingma, *CIBIT Consultants | Educators*, Arthur van Schendelstraat 570 P.O. box 19210, 3501 DE Utrecht, the Netherlands, tel. +31-30-2308900; fax 31-30-2308999; e-mail: rvdspek@cibit.nl, jkingma@cibit.nl; internet www.cibit.nl

[2] Hofer-Alfeis, Josef, van der Spek, Rob: Knowledge Strategy Process and Metrics. In: Proceedings of the Knowledge Management & Organisational Learning Conference, 12-15 March 2001, London, Linkage International Ltd., www.linkageinc.com, p. 153-176

[3] Hofer-Alfeis, Josef, et al: Strategic Management of the Knowledge Enterprise. In: Proceedings of the 6th APQC KM Conference "Next Generation Knowledge Management: Enabling Business Processes", 10-11 Sept. 2001, Houston, Texas, APQC, www.apqc.org

[4] Zobel, Joachim: "Knowledge Management bei PriceWaterhouseCoopers Deutschland" (Knowledge Management at PriceWaterhouseCoopers Germany); presentation of the Deutsche Gesellschaft für Personalführung (German Society for Human Resources Management), 25 Jan. 2001, slide 36

[5] TECTEM, Benchmarking Center, Universität St. Gallen, Switzerland: Benchmarking Project Knowledge Management, Screening Report, 2000, p. 37

[6] Boisot, Max H.: Managing Knowledge Assets – Securing competitive advantage in the information economy. New York: Oxford University Press, 1998

Key propositions

1. The Knowledge Strategy Process is a dedicated instrument used by business owners and their teams to strategically plan the use and management of business-relevant knowledge from their perspective.

2. It identifies which knowledge areas have an impact on the business, how strong this impact is, which deficits there are in each of the knowledge areas in terms of proficiency, codification and diffusion, and what the management feels it can do in response to these issues.

3. A major output of a business-specific Knowledge Strategy is the actions for various knowledge-related management disciplines strengthened by the buy-in of the management team. One group of actions is of cross-business importance, i.e. this group solves common issues in many knowledge areas and in different businesses. These actions provide a valuable input of requirements for the Chief Knowledge Officer's KM roadmap. Another group of actions basically concerns business-specific knowledge issues and has to be implemented by the related managers in this business.

4. The resulting projects are then drafted and implemented – with the cooperation of an inter-disciplinary knowledge management team and a KM consultant, if required. Thus orchestrated and state-of-the-art KM solutions can be achieved.

5. Success can be directly monitored by reviewing the various specified actions, objectives and impact relationships in the knowledge strategy.

6. The Knowledge Strategy Process (KSP) should be integrated into the business strategy process and revisited regularly.

7. The KSP has been tested within Siemens AG using practical applications in two pilots and several training sessions for KSP consulting.

Discussion questions

1. What are the main reasons for Siemens AG co-developing and applying the Knowledge Strategy Process?
2. What is the role of key performance indicators in the KSP?
3. Which participants are involved within the KSP and what is their contribution?
4. What are the three main results of a KSP?
5. What is the difference between a Knowledge Strategy and a KM roadmap or KM strategy?
6. How is the buy-in of the business owner and his management team to knowledge-related management actions achieved?
7. How can you ensure that the KSP is integrated into and aligned with the business planning cycle?

II Knowledge Transfer

ShareNet – the next generation knowledge management

Michael Gibbert, Stefan Jenzowsky, Claudia Jonczyk,
Michael Thiel & Sven Völpel

Abstract

Leveraging knowledge on a global basis is a major challenge of big multinationals like Siemens. Induced by significant changes within the international telecommunication business, Siemens Information & Communication Networks (ICN) faced a shift in competitive pressures that stressed the necessity for knowledge-based competition. The ShareNet case describes how Siemens ICN succeeded in its transformation from mainly a product seller to that of a global solution provider. It outlines the role ShareNet, a global knowledge management network, played within this transformation and discusses the critical success factors involved.

Background

Kuala Lumpur, Friday afternoon. Two intensive weeks of hard work awaited Martin Wong. As the Manager at Siemens ICN Malaysia, he was responsible for the telecommunications business with Malaysia Telecom, one of Siemens' most important Asian clients. Martin needed to complete a comprehensive proposal for a voice-over IP network solution for Malaysia Telecom within two weeks. This was the first proposal of its kind for Siemens Malaysia's business unit.

While working on this proposal, Martin had to come up with answers to questions like:

- Which technical solution would suit this situation the best?
- Should Siemens immediately offer an existing service package?
- How exactly could he demonstrate to this specific customer that, in a very competitive environment, the Siemens solution would best fit his needs?
- Where could he get hold of a business plan, at short notice, that would show the customer how soon the Siemens solution would be profitable?

In the past, finding these answers alone might have taken him many weeks, or even months.

Today, the answers are just a mouse-click away for an expert salesman like Martin Wong: His company's intranet offers him access to the Siemens ICN ShareNet – a global knowledge database that provides him answers to those tough questions. In Share-Net he will find similar customer solutions with their accompanying sales arguments, descriptions of successful projects, presentations, relevant business plan, as well as several contact persons who could help him with questions on technical issues, or financial concepts. The crucial proposal can be compiled quite quickly and Martin will be able to focus on his core competence – developing strategic solutions with the customer.

ShareNet is an example of how practical knowledge management within Siemens has had a substantial effect on its business success. ShareNet links the salespeople of Siemens Information and Communication Networks (ICN) worldwide, making each salesperson's accumulated learning experiences accessible to the entire sales force. This facilitates sales, helps to save valuable time and money, and leads to increased revenue with higher profit margins.

With the telecommunication industry's strategic context characterized by great flux (as described later) the codifying and sharing of relevant knowledge through database-media has become much more difficult. Recognizing the risk of being saddled with codified knowledge in obsolete data graveyards, ICN ShareNet went beyond the mere hoarding of information in data repositories. It focused on orchestrating an interactive web of knowledge and expert networks on a global scale.

The shifting context in telecommunications

The changing landscape in telecommunications

From the inception of the telephone service until the 1980s, customers of telecommunication equipment around the world had mostly been of one type: the monolithic, integrated telephone company. The entire range of activities involved in providing telephone and data services to the end-user, i.e. the entire value chain, starting with the planning of the network to its implementation (including customer acquisition and care), was concentrated in a single, large entity. Being less cost-sensitive by nature these telecommunication monopolies usually focused on long-term business relations with just a few telecommunications suppliers. The integration of the telecommunication equipment was normally handled by the monopolists themselves.

Previously the main business of a telecommunications-equipment supplier, such as Siemens ICN, was to manage the long-term relationship with its customer and to supply a range of well-engineered equipment. Time-to-market and pricing were of secondary concern. Consequently, the telecommunication-equipment suppliers of the past came to mirror their monopolistic customers: They also became vertically integrated, less sensitive to costs and oriented to the needs of a few, stable customers. Decision-making was centralized which, in turn, resulted in the flow of information following suit.

Times changed. Over the past two decades, governments worldwide have been deregulating the telecommunications-services market to provide consumers and end-users with more competitive pricing and better service. This led to the various telecommunication markets being at a different stage of their economic development. To complicate matters, technological advances in electronics and computer science led to an explosion of new products and services in the telecommunication services market. The end-result was a previously unknown diversity of telecommunication demands from all over the world.

Another consequence of this worldwide deregulation of telecommunication was the unbundling of the integrated, monolithic telephone companies. The large few were replaced by a variety of companies, often offering services in specialized market segments, such as telecommunication to certain foreign destinations, or specifically to business customers. The new, competitive landscape also led to the disintegration of traditional value chains. Once it was possible for a company to shift costs between services, for example by charging much more for long-distance calls that actually cost very little to supply, and using the margins on this lucrative service to subsidize residential services. Today, competing long-distance service companies, with no residential business to subsidize, can provide that same service much cheaper. Cost shifting is no longer possible.

The change in the telecommunication industry led to a radical change in the nature of the telecommunications-equipment business. Siemens ICN, a leading telecommunication-equipment provider, active in over 160 countries with 60,000 employees and a revenue of US$ 13 billion, served a variety of customers with very different needs. The

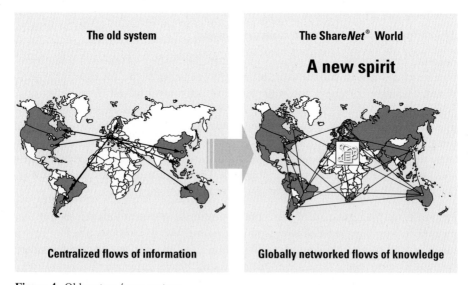

Figure 4 Old system / new system

plex solutions, ICN developed a practical approach that leveraged what had developed into a key factor in competitiveness in the new telecommunications landscape – sales knowledge and innovation. In order to stimulate and encourage empowerment, creativity, and innovation, Dr. Koch assigned the department Business Transformation Partners (BTP) the challenge of developing, rolling out and monitoring ICN ShareNet.

ShareNet is an interactive knowledge-management tool through which a global network of shared knowledge could be established. It was developed in close co-operation with the ICN board members, with Joachim Doering, the head of BTP and ICN Vice President, actively promoting the initiative in the different local companies.

The initial development of ShareNet

To ensure that ShareNet would be relevant to the day-to-day work of the sales people, the first step was to assemble a selection of the company's most successful sales people in a hands-on, knowledge-mapping process. Members of this core ShareNet team included sales representatives and local company heads from markets around the world, covering the full spectrum of business situations faced by the company. The question that this team addressed was "How do we sell solutions?"

The team developed a map of the solutions-selling process and identified broad categories of business-relevant knowledge for each aspect of this process. This rigorous approach also helped the sales people to realize how much they had to learn from one

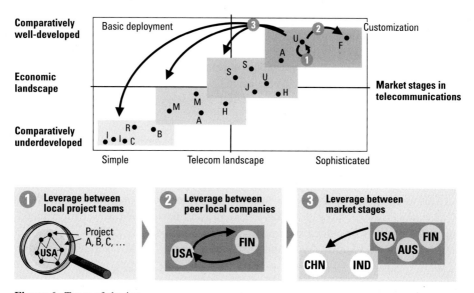

Figure 6 Types of sharing

another. A key insight gained through this mapping process was that not only the software and hardware building blocks of different solutions, but virtually every activity enabling a telecommunication service for the end-user, constituted a potential solution element that could be leveraged and re-deployed.

It soon became clear that knowledge sharing between the local project teams within a country – focusing on the same market, facing the same competitors, and therefore challenged by the same problems – could lead to a substantial competitive advantage. This type of knowledge sharing is called *leverage between local project teams (process 1).*

However, a fundamental question still remained to be answered: What would be the benefit of leveraging knowledge globally? Telecommunications markets were in different stages of development, leading to differing demands in these markets. The market stage depended on a country's economic development, as well as on the development of the telecom landscape.

Each country could be positioned on a two-dimensional graph by determining:

- its economic development – ranging from comparatively underdeveloped to comparatively well-developed – by means of its GNP
- its telecom landscape development – ranging from a simple landscape to a sophisticated one – by means of most important influencing factor, namely the degree of deregulation in the market.

Based on its position on this graph, a country's market stage in telecommunications could be determined. The market stage, in turn, determined the kinds of solutions demanded in the market.

Why then should solutions be leveraged across countries? The mapping of the sales process suggested that countries in the same market stage often addressed similar needs and therefore tended to seek similar solutions. By the same token, evolving telecommunication markets often encountered problems or upgrade pressures engendered by their more demanding end-users. These problems and upgrade pressures, again, tended to be similar to those previously encountered by markets that had now evolved to a more sophisticated stage.

This suggested that a solution sold in one country could be leveraged to another country at the same market stage, forming a so-called peer group. This type of knowledge sharing is called *leverage between local peer companies (process 2).*

As markets develop, solutions of the next market stage become more and more relevant to customers' market success. To allow customers to develop ahead of their competition, Siemens ICN leveraged solutions of higher market stages to those of the lower stages. This type of knowledge sharing is called *leverage between market stages (process 3).*

These three types of knowledge sharing do not require three types of systems. Leveraging between local project teams in itself leads to a significant competitive advantage, therefore installing a system to allow this kind of sharing should be profitable. If the same kind of system were installed all over the world, the system's interoperability would be guaranteed and knowledge would be reused. This would not only enable knowledge sharing between local project teams but also knowledge sharing between peer local companies and between market stages. By utilizing a single worldwide tool for knowledge sharing within one country (process 1), two additional byproducts are also obtained, virtually gratis – knowledge sharing between peer countries (process 2) and between market stages (process 3) – making it a very attractive prospect.

ShareNet – a business application system

The knowledge management initiative ShareNet was launched early in 1999 to provide sales people, worldwide, with relevant knowledge about solutions and applications, sales processes, and projects. With its aim to leverage knowledge and innovation globally, it was explicitly designed to foster the emergence of best-practice sharing, thus enabling a powerful learning process.

In this context, ShareNet nurtures the changed role of local Siemens companies throughout the world. In the face of new customer demands, these local companies evolve from mere outlets to companies with full responsibility for customer management. At the local-company level, the goal is to detect local innovations and leverage them on a global scale.

ShareNet avoids the problem of too great an emphasis being placed on information technology at the expense of in-depth business understanding that has proved to be a pitfall of many similar knowledge-management systems. Unlike traditional, often intranet-based, knowledge-management systems that have primarily been conceived as "document repositories", ShareNet provides a network that has been explicitly designed as an interactive medium. Instead of functioning as an infrastructure that exists alongside people's actual work, ShareNet functions as a business application, seamlessly dovetailing with employees' ways of solving customer problems. It covers both the explicit and tacit knowledge of the sales value-creation process, including project know-how, technical- and functional-solution components, and knowledge about the business environment (e.g. customer, competitor, market, technology, and partner knowledge). The emphasis here is on experience-based knowledge. As shown in Figure 7, knowledge about the different steps of the value-creation chain was transferred to ShareNet solution objects (e.g. technical- or functional-solution knowledge) and ShareNet environment objects (e.g. customer or market knowledge). ShareNet's focus is less on "brochureware", than on personal statements, comments, the "field experience" of sales employees, or the real-life tested pros and cons of a solution.

In addition to structured "questionnaires" on the above-mentioned topics, ShareNet provides less structured spaces, such as chat rooms, community news, discussion groups on special issues, and so called urgent requests. Urgent requests are basically forums for asking all kinds of urgent questions, such as, "My customer needs a busi-

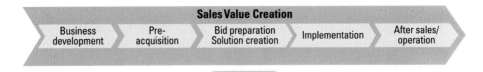

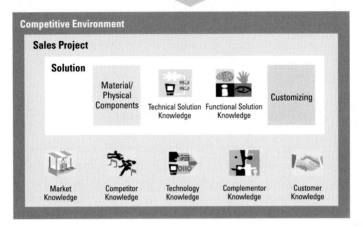

Figure 7 Value creation chain

ness case for implementing the new technology X by next Monday. Who can help me?" or "Does anybody have a list of recent network projects by competitor Y?" These are, in other words, questions that do not have a defined organizational owner. As ShareNet works independently of time zones and organizational boundaries, members usually receive answers within hours. In many cases, the right answers are "harvested" and made available for later use in a FAQ (Frequently Asked Questions) section. Thus, unlike traditional knowledge-management systems, ShareNet is based on an interactive approach and mobilizing knowledge and innovation in sales.

Mobilizing global knowledge sharing

Identifying areas of intervention

An important concern in the development of ShareNet was the adequate positioning of initiative as a true value-adder that helps to solve relevant problems in employees' day-to-day work. It was critical to emphasize this to prevent ShareNet from being portrayed as yet another headquarters project that would be demanding precious resources. This was the goal from the onset. It started with ShareNet's development as a joint effort of a core team of sales people from all over the world who recognized that local sales and marketing people felt that they too had a vested interest in the development of such a system. This was mainly achieved by addressing four interrelated areas of intervention.

1. Cognitive knowledge – or know-what – is defined as basic technical mastery and is achieved through extensive training and certification. For ShareNet this means technical knowledge, for example in the form of pricing concepts, represents an essential, but not complete, aspect to ensure commercial viability.

2. Skills – or know-how – refers to the effective execution and application of abstract rules and regulations in the real-world context. ShareNet achieves this through the feedback given by sales professionals in de-briefing projects.

3. Systems understanding – or know-why – refers to a deep understanding of cause-and-effect-relationships underlying an experience. In a global-sales-and-marketing context, this enables professionals to anticipate subtle aspects in their interaction with a customer. This understanding is especially important in view of the increased complexity of the sales process. For example, an experienced key account manager will instinctively know which components of a solution can be developed further, be leveraged and re-deployed in other countries, or even re-invented to suit different requirements. The Systems understanding therefore represents a particularly important area of intervention.

4. Self-motivated creativity – or care-why – refers to an active and caring involvement in a given cause. For ShareNet this means systematically identifying and promoting highly motivated and creative groups of employees. Indeed, such groups often out-perform other groups with greater resources.

These four areas of intervention together ensured that a user-friendly, accessible tool with authentic added value was developed for the sales and marketing staff. In the words of a senior key account manager:

"Offering a user-friendly tool, which can be accessed via the Intranet is not enough. You have to care for the people who are actually using it. You need a deep understanding of their ways of doing business and the problems they encounter. Ultimately, this ensures that you get the right attention and co-operation".

Critical success factors for global knowledge sharing

Designing a user-friendly tool was one thing but what it would look like in practice was another. In order to make knowledge sharing happen on a world-wide level, potential barriers obstructing the free flow of knowledge within Siemens ICN had to be anticipated and systematically eliminated. Joachim Doering and his ShareNet team identified five critical success factors that had to be considered, namely (1) leadership, (2) organizational structure, (3) motivation and reward systems, (4) organizational culture, and (5) a viable business case.

Leadership

Perhaps the most important critical success factor to making global knowledge sharing happen is the unconditional support of top management. In the words of Roland Koch, CEO of Siemens ICN:

"ShareNet is about collaboration beyond all existing organizational barriers. Our future lies in the creation of a net of knowledge spanning between all our employees".

Top management's support enhanced the value and strategic quality of the knowledge-management initiative and sent a signal to channel organizational resources and individual commitment towards this element. Management helped to communicate the idea of ShareNet across organizational levels and functional departments to ensure its added value was understood and appreciated.

The responsibility for the ShareNet initiative was given to the ShareNet Committee, the highest decision-making body of the unit. It was responsible for the strategic development of ShareNet worldwide. The committee was composed of eleven members: One member served on the ICN board, two members came from ICN Business Transformation Partners BTP, but the majority of the members were local company representatives. This guaranteed that the opinions of the local users of ShareNet would be heard and that they would be actively involved in the initiative. The size of the committee was deliberately kept small to enable its members to develop consistent decision-making competency and to react quickly to stimuli and suggestions from the field.

Organizational structure and rollout

The concept of ShareNet is probably more concerned about the managerial system and processes than about the technical platform itself. These managerial processes have been managed carefully from the first emergence of ShareNet. They cover the input of valuable knowledge as well as the elevation of this knowledge to more reusable (and thereby: abstract) knowledge. This task of making the knowledge inside the ShareNet system richer and more general and reusable is the prime task of the content editors, which are part of the ShareNet organization. This organization also contains country-specific consultants, IT-support, and a telephone and email hotline, providing answers and help for all users worldwide.

The ShareNet committee is the highest decision body of ShareNet. It consists of several heads of the Siemens ICN local companies and a few high-ranking headquarter managers, including the CEO himself. This committee was of utmost importance for the ShareNet rollout – which demanded devoting resources from all local companies. Local ShareNet managers are these resources. They are facilitators and trainers, ensuring the roll-out and support in their respective countries. Without the network of the ShareNet managers, the rollout of ShareNet would not have been possible. The Share-Net managers were trained, prepared, and outfitted for their task at one worldwide ShareNet bootcamp, organized by the ICN Business Transformation Partners organization.

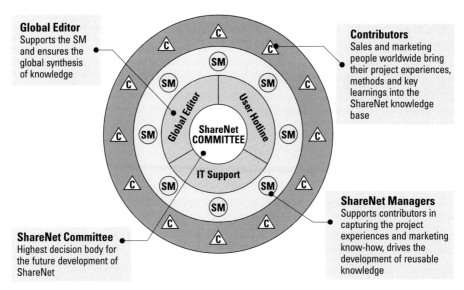

Global Editor
Supports the SM and ensures the global synthesis of knowledge

Contributors
Sales and marketing people worldwide bring their project experiences, methods and key learnings into the ShareNet knowledge base

ShareNet Managers
Supports contributors in capturing the project experiences and marketing know-how, drives the development of reusable knowledge

ShareNet Committee
Highest decision body for the future development of ShareNet

Figure 8 The ShareNet organization

The backbone of the ShareNet organization are – of course – the users, which are contributors at the same time. They are the main driving force behind the system, and they form the network of friends and colleagues that make knowledge management work for Siemens ICN.

While technology can certainly act as a facilitator for global knowledge creation and sharing, especially in the case of explicit knowledge, it is erroneous to believe that high-volume, quantitative data repositories can significantly improve organizational knowledge assets. Since knowledge is not static, but subject to continuous modification, it cannot be frozen into depositories. In recognition of this, ShareNet had to ensure adequate levels of interactivity in order to conserve the dynamic nature of knowledge.

To make knowledge sharing happen, interactivity was also required on an inter-departmental, inter-divisional, and inter-functional level. It is often difficult to accept and adopt another person's knowledge, especially if this person is from another division or department. An account manager at ICN commented on this "not invented here" syndrome:

"Sometimes knowledge, which has been brought in from external sources, such as another Siemens departments or divisions, raises defense reactions. People often do not use it for the simple and stupid reason that they did not invent it. We have to develop people who can integrate suggestions from different origins and make a successful project out of it. In short, make things happen, even if a project is composed of external inputs only".

Motivation and reward system

It was necessary to systematically identify and eliminate any organizational structures that could prevent knowledge from being shared, leveraged, and enriched by different functions and departments – and across organizational levels. A critical success factor, therefore, was firstly the establishment of a targeting and compensation system for top managers (see the "Bonus on Top Case").

But a targeting and compensation system for top managers was not enough. On the level of the employees actually using ShareNet, a motivation and reward system was developed that removed the fears and anxieties that could prevent the exchange of knowledge across divisions and departments. Knowledge in general, and sales knowledge in particular, is bound to a person. It cannot be shared with others against a person's will. This raises questions about motivating people to share their knowledge. Getting a person to enhance other people's knowledge by voluntarily contributing his or her own does not happen easily. A further constraint is that it is considered a time-consuming and tedious exercise. In fact, the individual contributor might wonder how he or she could possibly benefit. An important benefit for the individual contributor is to portray the individual concerned as an expert in a certain field. The drawback is that once this reputation had been gained, others may often solicit this expert's opinion, leading to time lost for the individual's own projects.

The need to motivate and reward such sharing is equally important for both the contributor (or "giver of knowledge"), and the re-user (or "taker of knowledge"). The contributor, who receives no direct reward for making experiences available, has to be specifically rewarded for the time invested in sharing his or her knowledge. The main reward for the re-user is the knowledge itself, which facilitates daily work. Yet, rewarding individual performance can lead to another counterproductive result. During the ShareNet implementation people were reluctant to adopt knowledge from others. The "not invented here"-syndrome described in the organizational structure, is closely related to this. The willingness to re-use existing knowledge became crucial for this initiative to fully succeed.

For the re-user to benefit and thus gain the reward, ShareNet had to ensure that the available knowledge was truly useful. This was done through stringent quality control. Nevertheless, a reward beyond that of gaining knowledge, significantly improved the re-user's motivation to re-use knowledge. The ICN ShareNet Quality Assurance and Reward System is designed analogous to frequent flyer mile systems found in the airline industry. As shown in Figure 9, contributing and re-using knowledge is rewarded by ShareNet "shares". Depending on the number of shares accumulated during a year, employees are awarded with several incentives, such as conference participation or telecommunication equipment. The number of shares given to the contributor depends on the re-use feedback of the taker of knowledge, thus rewarding the usefulness of the transferred knowledge. The higher the usefulness of the knowledge, the higher the reward is. The feedback mechanism is an important part of the quality-assurance system, too. The quality of available knowledge can be quantified through re-use feedback from several knowledge re-users. Based on this feedback, knowledge of an inferior

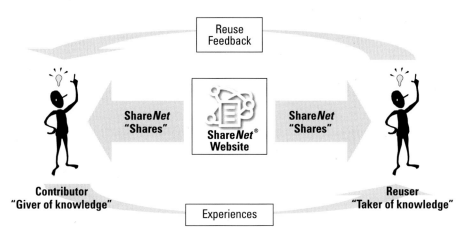

Figure 9 Reward system

quality can be removed from ICN ShareNet, whereas high-quality knowledge can be identified and developed further. This leads to a constant improvement of the quality of the available knowledge.

Organizational culture

Organizational culture as a set of beliefs, attitudes, and assumptions is mainly concerned with the unwritten, less visible part of the organization. Symbols, ceremonies, office settings, and dress code are examples of organizational culture. Additionally, it determines the way in which people interact and work together, and also prescribes rules and regulations about what is considered acceptable.

Organizational culture has vast implications for the implementation of knowledge management at Siemens ICN. To a large extent, knowledge sharing depends on the quality of the relationship between employees, as well as their relationship with management. A culture of openness, mutual respect and the absence of ambiguity is fundamental for fostering knowledge sharing.

A strong hierarchy often counteracts such an atmosphere since it promotes individual performance at the expense of team performance. Promoters of ShareNet, like Joachim Doering, worked hard to spread the ShareNet message that "unlike in school, copying is not only allowed – it is required". Another barrier was that the strong hierarchy naturally directed responsibility towards the top, whereas a culture conducive to knowledge sharing is built on empowerment.

A viable business case

A viable business case was a key factor for a successful knowledge-sharing project. The IT system, the motivation and reward system, the change of organizational structure and culture all contributed to making ICN ShareNet expensive. ShareNet, therefore, had to illustrate its benefits with a realistic business case.

Of course a knowledge-sharing system is expensive – but so is the continual labour of rediscovering solutions. The costs of sharing knowledge are quite obvious, the benefits are less so. There are three types of somewhat quantifiable ShareNet benefits:

- The saving of costs, e.g. by re-using tenders or re-using knowledge on how to simplify processes.
- Increased revenues, e.g. by increasing the quality of tenders by re-using knowledge of the success factors of tenders, or by simply being faster than the competition by re-using documents.
- The alignment with customer needs, by recognizing important trends and developments worldwide.

Perspectives

Since the beginning of this millennium, ICN ShareNet has become an integral part of the strategy of Siemens ICN. Dr. Koch, CEO of Siemens ICN, remarked:

"This [ShareNet] network will be of key importance to the success of ICN's solutions business because the company that can make use of existing experiences and competencies quickest has a distinct competitive edge over other players. We need to be among the first to realize this strategic competitive advantage through efficient knowledge management".

ShareNet can be improved further. With the community of 7,000 sales, marketing, and business-development people at Siemens ICN worldwide who actually comprise ShareNet, the use of ShareNet has reached a critical mass. Within its first year of existence, it has developed into a tool of practical knowledge management, enabling improved sales and marketing processes, faster action in the marketplace, and knowledge-based competition. The ambitious target of earning 250 million Euro in additional revenue in the first year of ShareNet's implementation has still to be met.

Joachim Doering, vice president of Siemens ICN, believes that ShareNet has an even greater potential to realize a measurable business impact through the creation of new business opportunities. As a next step, new communities, such as the worldwide service units and R&D, have to "come on board" to develop ShareNet into a knowledge portal that integrates the expertise of the whole enterprise in virtual workspaces.

Broadening the focus of ShareNet internally to include other functions is not the only task ahead. Joachim has a clear vision of what the next steps should be. He envisages expanding ShareNet across organizational boundaries to integrate customer knowledge into the system. In this context, new questions arise, such as, How can customers be motivated to participate in the ShareNet initiative? What exactly is the critical knowledge ICN expects to gain from its customers? And last, but not least, the broadening of ShareNet across ICN boundaries gives rise to a whole range of completely new issues, such as security and confidentiality concerns.

Finding the answers to these questions is the key to leveraging the potential of Share-Net in particular, and Siemens ICN's future, in general.

Kuala Lumpur, Two weeks later. Martin Wong moves his chair back with a sigh of satisfaction. It has been an extremely long two weeks. ShareNet has not left him twiddling his thumbs, but it has made his job so much easier. After receiving input from around the globe, he is certain that the proposal which he has just completed will sweep the opposition aside. This weekend he needs to reward himself! Now, what would he like to do....?

Conclusive remark

Because of the high interest the ShareNet concept affectuated at the Siemens divisions and outside Siemens, a startup company called The Agilience Group headed by the former Siemens manager Dr Christian Kurtzke was chosen to leverage the key success factors of ShareNet to the world outside Siemens ICN.

Key propositions

1. The ShareNet case demonstrates the importance of finding the right balance between IT solutions for capturing explicit codified knowledge and leaving enough room to allow direct personal exchange of more implicit forms of knowledge.
2. Knowledge-management initiatives have to be embedded within appropriate incentive systems, structural arrangements that facilitate knowledge sharing and an organizational culture that supports such an initiative.
3. To ensure the global reach of the knowledge-management initiative, knowledge sharing has to take place on three levels: within one country, between peer countries and between market stages.
4. The two components, "economic development of a country" and "degree of deregulation of the telecommunication market," determine the kinds of solutions that can be leveraged across countries.

Discussion questions

1. What were the driving forces that lead Siemens ICN to envision their becoming a solution provider and what role did knowledge play in this process?
2. What are the additional variables that have to be considered for global knowledge-management initiatives when compared with local initiatives?
3. What are the organizational factors that should be considered when designing and implementing a knowledge management initiative?
4. How should Siemens ICN tackle the future challenges ahead?
5. How can you motivate employees to share their knowledge?

SiemensIndustrialServices:
Turning know-how into results

Marc D'Oosterlinck, Hartmut Freitag & Joachim Graff

Abstract

This case study describes the Know-how Exchange, a unique knowledge-sharing network tool connecting the employees of Siemens Industrial Services. The Know-how Exchange is an extensive database compiled by those at the coalface of industry. They formed this innovative tool by pooling their knowledge, gained from practical experiences of diverse projects in about three hundred locations throughout the world, to channel it for use by others. The Know-how Exchange allows this invaluable knowledge to be shared and experiences to be reused, thereby increasing customer benefit and process performance. Expert knowledge, the fruit of many years' innovative work, is now transparent and available via each employee's individual PC. The immediacy and practical application of this tool are related via an authentic (but fictitiously related) narrative, interspersed with additional motivational and background information of its development, problems and current state.

The following situation is authentic – however names and places have been changed, where necessary.

India, April 2000

"Seventy-seven million Euros"! Stunned, Rajiv Krishnan puts down his knife and fork. "Did I hear that correctly?"

His customer, and dinner companion, nods enthusiastically. "You got it right the first time! Our competitor is going to build a textile plant for 77 million Euros in Tamil Nadu. That's practically next door to the Siemens office in Cuddalore. It is being said that the new plant will employ seven hundred people and will serve the US and European markets. A German company in Frankfurt, Schmidt-Meyer GmbH, has been awarded the planning and construction contract".

Rajiv conceals his excitement, but his thoughts are busy. It has to be possible for Siemens India to get at least part of the project.

Rajiv has been employed by Siemens Industrial Services in Cuddalore for eight years. He likes working there, has wonderful colleagues, a reasonable salary, and the work … well, it's work. Sometimes it is interesting, but tough when orders pile-up, and sometimes, when things slow down, he just cruises along. Clearly, as a sales person, he is always eager to get new orders. Unfortunately, he hasn't yet been successful with this particular customer.

It must be possible to find out whether the Frankfurt-based company is looking for a partner in India: a local partner with a global presence; a partner which, in similar projects, has proven competent and reliable; a partner who could supply the electro technical equipment and related services, like the installation, commissioning and maintenance of the plant – a partner like Siemens Industrial Services!

The next morning while making his way to the office by bus, Rajiv is still deep in thought. He has to find somebody who is familiar with the textile industry in Germany and who knows this company in Frankfurt; someone who could make contact with the company, and the sooner the better. The project has already been announced and that means that the development is at an advanced stage. A few days, or even hours, could make a major difference.

He is so deep in thought that he almost misses his stop.

In the elevator, he decides to search Siemens' global telephone book first. Unfortunately, this would mean losing approximately three and half days before he finally gets hold of the right person, since the telephone book does not provide any information about the listed people's expertise. Even worse, this person could quite conceivably be on a business trip. Wouldn't it be great if there were something like the Yellow Pages – only a lot better? It should include additional information, like the person's fields of expertise, experience and contacts.

The Turbo Yellow Pages

By the time the elevator reaches the sixth floor, Rajiv has the answer: The quickest and best way of finding a contact person is through the Know-how Exchange. Soon after his colleague, Naresh Acharya, returned from the Siemens Industrial Services Asia conference, he told him about the Exchange tool – a sort of turbo version of the Yellow Pages. The elevator stops at the seventh floor, Rajiv hurries to his office and gets going. He starts the computer and a few minutes later clicks into the Siemens Intranet, goes to the home page of Siemens Industrial Services and from there, via a link to the Know-how Exchange.

Here he encounters the first hurdle: he needs to login. At this moment, this is really a nuisance, but it makes sense to him. After all, the company's internal knowledge should only be accessible to people who will apply it for the company's benefit and not to every casual web surfer. Rajiv glances at the clock: 09:30 in India and 06:00 in Germany. It's far too early to call, but it does mean that he has plenty of time to send an

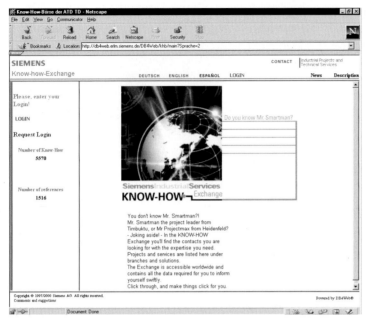

Figure 10 The homepage of the Know-how Exchange

urgent email message to the head office in Erlangen, Germany, requesting a login and password. In the interim, he can read the description of the Know-how Exchange online. This will bring him up to speed quickly, allowing him to start his search as soon as he receives the login and password.

Three hours later his request is answered. That's pretty good going when you consider the time difference, and now he need never apply for this access again. Rajiv logs on.

A simple, user-friendly screen opens. From here it is possible to access Siemens Industrial Services' know-how and overall competence worldwide. Rajiv knows this is only true insofar as individual Siemens employees have entered the relevant information. Currently the information available on-line is regarded as just a fraction of what is actually out there. Still, it is better than nothing and the Know-how Exchange is continually being updated and enlarged. He clicks on the "Know-how" hypertext link.

Siemens' quest for knowledge

The idea of a Know-how Exchange was born a good three years ago. Siemens Industrial Services' employees were tired of having to start from scratch with each new contract to build or modernize a plant, basically re-inventing the wheel each time in preparing quotations, contracts, project structures, engineering layouts etc. Not only was it

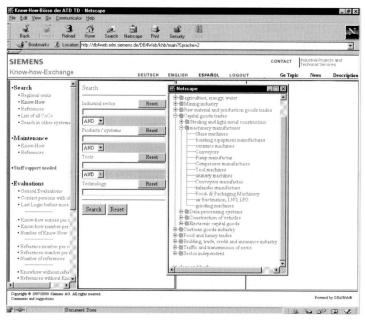

Figure 11 The main screen for know-how retrieval

expensive to start from the very beginning again each time, but it was also time-consuming and prone to pitfalls.

Most Siemens employees had never consciously thought about the immense amount of knowledge that the company had accumulated in the century and a half since its founding. The company's services cover a wide spectrum like engineering, integration, installation, commissioning, maintenance, repair and modernization of electrical and automation equipment as well as IT-systems for a large number of industries like power stations, breweries, dairies, textile plants and production lines for the automobile industry. In some cases, this know-how wasn't particularly valuable, because the pace of innovation had overtaken it. In other cases, however, it just wasn't possible to access experience and information from previous projects, as this experience base had either vanished when employees had left the company, or had ended up in dusty archives, neither of which were readily accessible.

Today, Siemens Industrial Services employs almost 22,000 people in more than 70 countries. Wherever electrical and electronic equipment are used industrially or commercially, Siemens Industrial Services are able to provide services like engineering, installation, maintenance and repair. Every day, each one of its employees gains new experience and learns new skills. Each thing that Siemens' employees have thought of, developed, systemized, built and checked, has generated know-how from which colleagues around the world can benefit. But powerful tools are needed to accomplish this transfer of know-how quickly and easily.

Creating the Know-how Exchange

Early in 1998, Dr Hartmut Freitag put together a small team in Erlangen that included Joachim Graff and Marc D'Oosterlinck. This team developed a database that can be searched for specific know-how and skills, and allows the user to introduce his or her know-how to others. It is a global Intranet-based database available in German, English and Spanish and, while the team had the system running after only a few months, there was one rather important drawback – no data!

Siemens Industrial Services employees were prodded, cajoled and motivated through newsletters, the Intranet and the employee newspaper, through personal emails and at congresses and conferences to get involved by contributing their know-how to this new tool. Every single employee, regardless of rank, had to be informed of the Know-how Exchange and learn what it could offer. Everyone in the company had to know that the success or failure of the database depended on him or her personally, because it would only be useful if it had sufficient data.

The emphasis on personal responsibility was necessary to prevent the initial euphoria changing into disappointment. One cannot find information that is not made available, and the database would only achieve break-even point when anyone looking for information could obtain it within a reasonable time-span. Furthermore, one successful search would definitely lead to further utilization of the database.

We are happy to report that we are now well beyond the break-even point.

But let's get back to Rajiv…

Back in India...

Rajiv accessed the textile industry on the Know-how Exchange by clicking first on the "References" hyperlink and then on "Industry sectors". The latter wasn't easy to find and he feels there is room for improvement here. It would be better, he thinks, to have an alphabetical list or to be able to search freely, instead of having a catalogue structure. He simply wants to know whether someone in the organization has been, or is involved in, building a textile plant in India. He has no luck. This does not mean that he concedes defeat. He feels sure there could still be someone – somewhere. It doesn't really matter, he was just curious any way. He does need a contact person in Germany though, as soon as possible. Ideally, it should be someone who has cooperated with this company in Frankfurt.

Rajiv does a search on know-how in the industry sectors and then, specifically, the textile industry. His perseverance is immediately rewarded: seven contact partners, arranged according to their geographical location. The first three are from Germany. He decides to investigate these three further. He clicks on a name and then the person's name, department, region, telephone, fax, email and Intranet address pop-up on the screen, followed by fields detailing the competence levels of their personnel. He ignores this plethora of information for the moment, deciding to examine it later.

He starts composing an email in which he can kill three birds with one stone.

"Dear colleagues in Germany,
A textile plant will be built here in Tamil Nadu. Schmidt-Meyer GmbH, based in Frank-
furt, is the project consultant. We would very much like a piece of the action! Does any-
body know the Siemens person who is looking after Schmidt-Meyer?

Regards,
Rajiv Krishnan from hot and steamy India".

A Siemens employee, based in Bayreuth, answers him just four hours later. Rajiv thinks hard. Bayreuth must be somewhere in Bavaria, a few kilometers north of Munich. She tells Rajiv that the Bayreuth employees had contacted Schmidt-Meyer immediately and found the responsible project manager. An appointment had already been made between the Bayreuth people and Schmidt-Meyer to discuss the project details the following day.

Things are moving fast. In spite of the air conditioning, Rajiv is perspiring, but really enthusiastic. His customer has already told him that the electrical portion comprises about 12 percent of the entire project. If only they could get this part of the contract! Siemens certainly has many experienced specialists in this field. Rajiv frowns and thinks hard. Yes, this calls for a case study.

He'll assume that, among other things, Siemens will supply the complete process auto-mation with a visualization system and all the peripherals. That would be realistic. Rajiv sighs. He doesn't have much experience with process automation and he's cer-tainly no specialist. This means that he urgently needs support from the worldwide know-how database, but will it work?

References are the best selling tool

Let's assume that Rajiv now wants to find out all the places where the automation equipment, SIMATIC S5 (a programmable logic controller manufactured by Siemens) has been used. This is possible, as the Know-how Exchange allows for a search for ref-erences. He clicks on "References" and enters SIMATIC S5 under "Products".

The result is 235 matches, ranging from water-treatment plants and high-bay racking systems, to the Tegernsee Bahn (Tegernsee cable railway). All these contact partners could tell him more about their specific projects and experience, besides confirming that Siemens knows more than enough about this technology and that's the trump card he needs. He can also provide details on which products, technologies and tools have been successfully used. The real test is always the specific application. The customer only accepts and pays for that which really performs the required functions.

Rajiv comes to the conclusion that it is actually strange that there aren't more refer-ences. What a great opportunity the Know-how Exchange offers employees to show that they are truly competent! He is inquisitive, however, and starts wondering what Siemens is doing worldwide.

Without applying any search filters in the form of specified products or tools, he browses through the whole industry sector. Here's a wealth of information indeed!

Siemens is providing process control technology for the Reichstag in Berlin (German Parliament building); building automation systems for a supermarket in Madrid; providing the electronic systems for a paper mill in Sweden; performing the electrical and mechanical installation for a Saudi Arabian project; building an office for the Deutsche Bank in Moscow and applying SIMATIC S5 for a Japanese company in Austria. Could there still be people who think that Siemens isn't a global player? In fact, the Austrians have even used SIMATIC HMI for an artificial snowmaking machine. Rajiv starts laughing. He wonders what the business potential of an artificial snowmaking machine would be in India!

"Are you coming, Rajiv? We're supposed to be lunching together. Remember?" ask two hungry-looking colleagues from the neighboring office.

"I'm on my way", promises Rajiv, as he takes a last glance at the computer screen where he has been looking for business partners. He has been unable to find an entry for Schmidt-Meyer. Oh, well, that would have been asking for too much.

Know-how sharing is important

On their way to the canteen, Rajiv tells his colleagues about his experience with the Know-how Exchange. Indira, another colleague who has joined them, is skeptical.

"How does one go about convincing people to download their know-how so that everybody can access it? People who are experts in any particular area like to keep their know-how to themselves. Who wants somebody else profiting from your experience, while you are relegated to the background?"

The men look at one another and shrug their shoulders. They are well past this stage.

"I see it differently", explains Rajiv. "Look, I am only going to put an appetizer on the Net – let's call it bait, for want of a better word. Anybody who wants to know what we can do, can take a nibble. We're not going to lose any know-how or experience that way. Any way, Indira, you simply can't download the real stuff. It should be possible, however, to download standard operations like planning, project management, or quality assurance as a module. I have already found some basic examples of this in the Know-how Exchange. We could save an enormous amount of money with those standard operations which are used almost everywhere".

But Indira isn't easily convinced. "Come on, Rajiv! The bottom line is that someone, somewhere, has created this know-how and it belongs to that person. If you can simply retrieve this know-how from the Net, it means that its owner, an experienced colleague, is no longer needed".

"I think differently", counters Rajiv. "The chances are good that I'll be contacted far more often than I am now if I am listed in the Know-how Exchange. In fact, I may even have to schedule appointments on the Net so that I'll have time for my normal work.

"You know, in future we may have to prove our productivity, and transferring knowledge is productive! Naturally, when doing so we must fulfill Siemens' corporate principles: "Our cooperation has no limits"; "Learning is the key to continuous improvement", and especially, "Customers govern our actions". In today's competitive global environment this means that competence is everything, but we must also be fast. Our customers benefit when we apply our know-how sensibly. We have to act more quickly, learn faster, and be more convincing when dealing with customers. That is the best basis for full order books".

They carry their trays to an empty table.

"Have you heard the story of the dollar?" asks Naresh. "They told this at the Asia conference. The story goes like this: If I give you a dollar and you give me a dollar, then we each have a dollar. But if I give you an idea and you give me an idea, we each have two ideas. This really makes sense to me. A dollar stays a dollar and doesn't increase in value even if I pass it on, but if I pass on an idea worth a dollar and discuss it with somebody else, I often receive a good tip. Then, all of a sudden, this idea is worth two dollars. The other person may also implement my idea and make a dollar fifty, or possibly even three dollars, but we have both benefited"!

A win-win situation for everybody

Indira takes a forkful of sesame rice. "Distributed know-how is therefore at least double the original know-how, right? But you must first feel confident distributing know-how, Naresh. The experts must have a closely-knit, readily accessible network. Everybody must know each other. Previously, when the units were small and transparent, that wasn't a problem. Everybody knew what everyone else did. Today, however, we are spread throughout the world and partners and responsibilities are in continuous flux, which means that it is extremely difficult to keep an overview, as well as work profitably, and that is what Siemens essentially aims to do".

"Okay, you convinced me. Let's get listed in the Know-how Exchange as soon as possible! Our colleagues worldwide should get to know us", Naresh responds. He sounds almost sarcastic.

At this point Rajiv comes in: "Corporate principles are not the only reason why we should be listed in the Exchange. Let's face it, there isn't a better marketing instrument. If somebody, somewhere, is searching in a specific industry sector, for tools or technologies for a specific project, and if I am listed in areas where I am competent and have experience, then they will always be able to locate me".

"Now you're talking"! says Naresh enthusiastically. "We could say that we built the airport in Indonesia. That wasn't at all as simple as it sounds. One of the problems was that the local power utility just didn't have the capacity that we required. It goes without saying that, in future, I will consider this in the planning phase – and others could learn from this experience".

It's Rajiv's turn to be skeptical. "I wouldn't want to state that I hadn't done my planning thoroughly enough".

Surprisingly, Indira doesn't have a problem with this. "Why not? Nobody is perfect and Naresh is not giving anything away by admitting mistakes – it isn't a sign of weakness. He is simply giving our colleagues a tip so that they don't fall into the same trap".

Naresh nods. "I'm assuming that nobody's going to take advantage of my reflecting on my mistakes, but that they would rather use my experience to expand their experience base. I am also assuming that my know-how won't be used against me if I make it generally available".

"Hey, if we can't assume that, then the Know-how Exchange wouldn't work at all", says Indira. "You learn through mistakes – your own, or those of others. In fact, I would like to list everything that hadn't worked out somewhere or other, but that would only work if the level of trust increases enormously. That's the most decisive issue for me when it comes to know-how management".

The project developers explain...

Such a list of mistakes does not, in fact, exist, but the Know-how Exchange user will find the names of the personnel who were involved in every project (those that have been completed and even some that are still running). These people would know everything about the project, including the difficulties. And you would only have to ask them to find out.

Back to a canteen in India...

In his mind's eye Rajiv can already see the completed textile plant. It's a great reference project: a state-of-the-art building with cutting-edge technology. Even the workers are proud to work there. The customer is satisfied with the services provided by Siemens Industrial Services.

"No matter what happens, I'm going to enter this in the Know-how Exchange when we're ready. Then I'll also have my reference list available anywhere and everywhere". Satisfied, Rajiv starts on his dessert.

Problems encountered with the Know-how Exchange

Initially there were all kinds of problems with the entries. This was not a unique experience, but a well-traveled road in software application. Siemens Austria wanted to enter all of their competences and references, since they had immediately seen the potential of such a knowledge database. No matter who the users were – browser, contact partner, or a successful supplier – everybody would profit.

The information was entered quickly and at first nothing seemed amiss. Then: ERROR! TIME OUT! The problem was finally traced to the server, or more specifically, the insufficient capacity of the line between Erlangen and Vienna. In all the excitement, nobody had thought about this. The consequences were painful and embarrassing.

Employees were disillusioned and frustrated. Happily, the problem has now been solved. The line capacity is sufficiently large and the system functions perfectly.

The people from the Know-how Exchange have identified essentially two types of users.

• The first group finds IT exciting and gets involved in any innovation quickly. This group finds difficulties a nuisance, but simultaneously a challenge. No problem can dampen their enthusiasm.

• The second group is somewhat more hesitant when it comes to new tools. They wonder whether they should invest valuable working time in something they don't quite understand and whose direct benefits cannot be immediately verified. They want to find precisely what they are looking for within three clicks of the mouse, or tend to regard the whole system as worthless. This group has to be convinced of the system's benefits on an individual basis – which is not simple with a tool whose success is based on the involvement of many users.

Providing a user-friendly tool

For the sake of this second group, access to the Know-how Exchange should be as easy as possible. Only when there is foolproof technology, and no special training is required for using the tool, the tool will be used. The long-term goal is to expand the access to a know-how portal for a Siemens Industrial Services. This means that the user will no longer have to use predefined keywords only, but will be able to search freely.

The following graph, showing the number of entries of references, tools, products, customers, industry sectors and technologies, clearly confirms that many Siemens employees are already convinced of the advantages of the Know-how Exchange. In recent

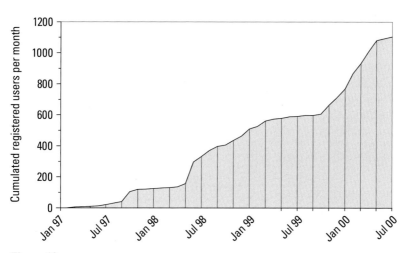

Figure 12
Interested employees can find over 1500 references and 5500 know-how entries, which are being added to every day

months the entries have grown by leaps and bounds, and the trend is continuing. To date, approximately 1,000 users across the globe have made entries.

The office in India...

On the way back to the office, the three colleagues decide to try the Know-how Exchange. At work they gather around Rajiv's desk. He not only has a login, but also the best air-conditioned office for the very hot Southern Indian afternoon.

Rajiv moves the mouse – three clicks and the screen for personal entries appears. He first enters his know-how and then the references. He knows that he has to trust that those who read his entries will not misuse his candor.

Communicating your know-how

Entries are completed in almost the same way as a search for data, except that the employee's qualifications must be entered as well. What do qualification levels Q1 to Q4 mean? A link provides the explanation. They denote competence levels, ranging from support and planning to specialist support and project management, from plant engineering and contract engineering to project administration and service, or administration with in-depth product know-how. It is important to work through these lists. It takes a little time, but assigning a Q level to the expertise of each employee provides a more precise description of the know-how available in the group. Individual names do not have to be entered.

The office, a little later...

"Yes, it really is possible to enter all the information from any computer – we don't even need any special archive. The reference list is always complete and updated". Indira, as always, appreciates a practical approach. "At least things aren't archived twice or even three times. I can always access all the data, even if a colleague's office is closed and locked, or if I am on a business trip".

Rajiv is still thinking about his textile plant. "When we generate a quotation, it would be great if, for example, we didn't have to re-calculate the cable length for each individual cabinet. There's a good chance that someone, somewhere, has already calculated it accurately. It must be documented somewhere".

Indira is absolutely sure about this too. "Of course it's written into some database somewhere. You only have to know where".

The project developers explain...

Siemens Industrial Services has the exact database the Indian colleagues are talking about. Unfortunately, it hasn't been linked to the Know-how Exchange yet, although the specialists are working flat out to achieve this. When this is ready, anybody will be able to generate a customer quotation from individual modules. This increases the planning reliability significantly and saves time. Many services are not specific to any particular industrial sector and can therefore be transferred easily. Once the link has been implemented, it will be possible to access a pool of standard solutions. For instance, this could be in the form of Excel tables which are best suited for a wide range of work. Using modules such as this will make quotations more accurate and reliable, because the details of the individual components have been tried and tested in practice. There will, no doubt, still be few unforeseen problems and issues, but there will also be more time to resolve them.

Clearly, the mere collecting of all interesting chunks of knowledge in a database is not sufficient. For a fast and reliable search with a good chance of success, a major prerequisite is providing simple, but powerful structures for the information. Everybody should be able to find what he or she is looking for quickly and easily. It should be equally easy for users to make their own know-how available. Even better, not only be able to make it available, but also to want to make it available.

In the office in India the knowledge adventure continues...

Rajiv has downloaded a set of overheads which can be used to present the Know-how Exchange. "To implement know-how and experience for the benefit of customers – that's it"! he explains, and shows the others an overhead. "Here you can clearly see that it is a win-win situation for everybody involved. Know-how, experience and skills can be distributed and discussed. You work with others and develop new ideas together".

Tom Whitley from Siemens Industrial Services in Chicago comes into the office. Tom is in Cuddalore for a couple of months. He sees the group around Rajiv's desk, comes closer and glances at the overhead. "Oh – the Know-how Exchange. What a great thing! I've already made some really useful contacts using it. But one day nothing would work. I suddenly realized that we had received a new version of Internet Explorer and the Know-how Exchange was optimized for Netscape. We were dumped right in the middle of the browser war. Luckily everything is up to speed again".

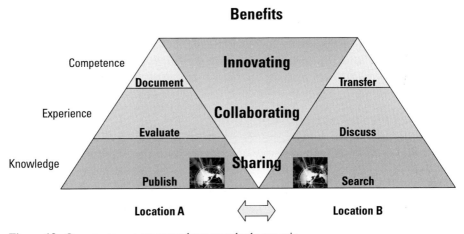

Figure 13 Our process – a process where everybody can win

"Yes", agrees Indira, "that's the kind of unexpected thing that sometimes crops up and suddenly you're confronted with a major problem. One can only hope that certain standard processes and systems will be established worldwide. Right? In fact, it is probably best to look for the "standard" in the Know-how Exchange before one starts a new task. This would help us to save time and optimize quality by reusing proven solutions. Practical standards are not only available for machines, but also for human behavior – we cannot stop progress".

Indira would even go so far as to enter that into the Know-how Exchange, thinks Rajiv. But she's right. You've got to be willing to distribute all types of knowledge and develop them. Everyone benefits from this, especially the person who is passing on the information.

On that note, it would now be appropriate for us to share with you, the reader, some of the knowledge we have accumulated during this project.

Key propositions

1. From this practical illustration of the Know-how Exchange, it should be obvious that the developers consider this project to be not only a success, but also one that has the potential to become a virtual Center of Excellence, available to every employee at Siemens Industrial Services, and possibly even further afield.

2. We realized that a tool is only a tool and its efficacy and success depend, to a large extent, on the people who use it. This aspect clearly poses a greater challenge to the developers than the technical side of things. Motivating people to make their expertise freely available to their colleagues for the benefit of Siemens and their customers, is no easy task, and one that will require ongoing attention.

3. By connecting geographically distributed service offices, the know-how-transfer process, supported by the Know-How-Exchange, provides an important competitive advantage and therefore plays a major role in the strategy of Siemens Industrial Services.

4. In order to fully exploit the potential of the Know-How-Exchange, know-how transfer has to be integrated into standard business processes. At any point in a process an employee should be able to learn from the experiences of his colleagues, as well as to provide his own experiences as lessons from which others can learn.

5. An area that certainly warrants attention now, and will do so increasingly in future, is the standardizing and structuring of the knowledge shared on the database. A certain level of knowledge quality is necessary to ensure its utility. Who will perform this gate-keeping task and what criteria should be used? These are questions that must still be answered if this tool is going to realize its full potential.

6. The introduction of a specific tool for know-how transfer should not impose an additional barrier for the users. A user-friendly tool and a reliable support team could avoid this.

Discussion Questions

1. When implementing a Know-how Exchange tool, what factors should be considered especially important?

2. Which barriers are most frequently encountered when it comes to know-how management?

3. What are the main benefits of the Know-how Exchange? Specify three points.

4. Would an employee who shares his know-how on the Exchange, be putting himself out of work sooner than an employee who does not share his know-how? Give reasons for your answer.

5. In your opinion, what is more important for know-how management: sophisticated software or human behavior? Discuss these two different aspects.

Networked knowledge – implementing a system for sharing technical tips and expertise

Andrea Dora, Michael Gibbert, Claudia Jonczyk & Uwe Trillitzsch

Abstract

In a sales and service environment, business is mostly driven by employees at the customer interface. By virtue of their intensive contact with the customer, sales representatives and service technicians are not only one of the major sources of knowledge and experience in these organizations, but also a source of information about future customer needs.

This case study considers the effect of a dynamic market environment on the opportunities for implementing a knowledge management program. After a short introduction to the organizational context, we discuss the market shift that created a need for the systematic management of knowledge. The study then gives an overview of a knowledge management initiative – its approach, relevant factors, and some of the underlying design principles.

A detailed description of KN Service Knowledge – an initiative for generating organizational knowledge for technical services – is provided, illustrating the activities involved in redesigning the work processes, IT architecture and collaboration with other departments. The human dimension is discussed, including the design of an incentive system used to bring about the necessary changes in behavior.

For the innovative and customer-oriented conduct that resulted from the repositioned linking of Knowledge Management and service, the knowledge networking initiative "KN Service Knowledge" was awarded the 2001 Service Management Award.

Introduction

Over the last ten years, the communication-network infrastructure business has undergone a change that has profoundly altered the rules for achieving and sustaining competitive dominance for Siemens Information Communications Networks, Sales Germany (ICN VD). Customer expectations have become increasingly more sophisticated with the simultaneous proliferation of products. With this growing complexity, companies have increasingly had to build up core intellectual and service capabilities to meet the requirements of the new environment. Value creation has become increasingly associated with developing highly individualized, knowledge-intensive solutions, rather than ones simply based on the sale of pre-packaged products. As a result, product experience has become less important. Siemens Information and Communication Networks (ICN VD) needed to take a proactive approach that emphasized the provision of complex packages of products and services, often referred to as "solutions". These solutions are typically developed in conjunction with, and specifically for, a single customer. As a result, they are highly individualized and demand substantial resources in terms of time and money.

Due to the geographic organization of ICN VD, complex customer solutions developed and provided in the Hamburg region, for instance, could not be simply reapplied in Munich. There was an urgent need for a more systematic management of the knowledge portfolio to prevent duplication in the provision of solutions. As the redevelopment of solutions from scratch means an enormous amount of time, work and also risk, the systematic sharing of localized knowledge across organizational boundaries became an urgently needed strategic lever to transform the business.

The emerging need for Knowledge Management

The organizational context

The setting for this case study is the German Sales and Services Organization within Siemens' business unit Information and Communication Networks (ICN). ICN is a specialized division that develops customized end-to-end solutions in the converging worlds of voice, data and mobile communications, for customers in industry, as well as those in the business and public sectors.

The Siemens' business segment Information & Communications (I&C) comprises Information and Communication Networks, Information and Communication Mobile and Siemens Business Services. Siemens I&C brings together the entire competence in the areas of networking technologies, telecommunications and information technology, and offers a whole range of products, solutions and services to customers worldwide.

Information and Communication Networks, Sales Germany (ICN VD), is the German market organization for the ICN business unit. With about 9,200 employees, ICN VD aims to provide diverse clients with solutions for data, mobile and telecommunications. In addition to selling products and services, its activities include consulting, professional services, technical services and training. It targets a customer base of more than 280,000 corporate and carrier clients in Germany and generates an annual turnover of around 2.6 billion Euro.

Like the whole Siemens Group, ICN VD has a decentralized organization. Several sales channels for specific market segments (e.g. carriers, small and medium-sized companies) serve the customers. Each sales channel is organized into sales regions, which, in turn, consist of individual sales offices within the region to ensure customer proximity. In total, ICN VD comprises six sales regions and about fifty sales offices, each of them enjoying a considerable measure of autonomy. The main business processes, such as Sales, Order Fulfillment, Marketing and Human Resources, are enclosed within one geographical unit.

This organizational structure, referred to as a geographic organization, also exerts an effect on the distribution and, therefore, in turn, the management of knowledge. As similar standard products were sold and serviced in all sales regions in the past, there was no need to exchange experiences and knowledge across regions. Within regional units, the sales and service representatives were accustomed to sharing knowledge in small networks that display the strong bonds of long-established contacts and personal friendships between a small group of employees. This localized approach worked very well while every geographic unit was doing the same business and possessed the same levels of competence. However, when business shifted towards more customized solutions and the variety of products, solutions, and services increased dramatically, the weaknesses of this method of sharing knowledge became apparent – it reduced the transparency of the internal knowledge practice.

Serving the customer as a solution provider

Top management at ICN VD pondered the implications of the shift in competitive emphasis and their strategic intent to transform the business from a "box mover", with its emphasis on products, to a "solution provider", which would focus on the provision of knowledge-intensive solutions. For Dieter Spangenberg, head of the business unit, Knowledge Management was a major key to adapting the organization to meet the new competitive realities:

The management of our knowledge assets constitutes not only an indispensable pillar of our business, but should be seen as the central element of our strategy as a solution provider.

Yet, this transformation meant that attitudes, behaviors and mindsets also had to change.

Practically every employee – whether in sales or services – possesses a rich portfolio of tacit knowledge and experience. This resource can only be utilized with the staff's active and voluntary collaboration. We need to get our colleagues to build a network of knowledge sharing.

The solution – a knowledge management program

Top management decided to implement a dedicated knowledge management program. A knowledge-networking team – comprising employees from support functions, sales representatives and service technicians – was set up to develop and run the program. The envisaged benefits included the re-use of proven solutions, reduced time-to-market and last, but not least, better customer services. Emphasis was placed on overcoming the limitations inherent in the geographic organization of ICN VD, and thus the development and implementation of tools and processes for knowledge sharing across individual regional boundaries. Hence, the interpretation of Knowledge Management at ICN VD became known as knowledge networking (KN). Dieter Spangenberg outlined the considerable breadth of knowledge networking in this statement: "The objective of knowledge networking is to create a living network of knowledge amongst all employees at ICN VD".

Knowledge networking

The goal and approach of knowledge networking

The most important aim of the knowledge management initiative at ICN VD was to provide support for the solution business. Due to the enormous complexity of this business, it was not possible for all the sales and service staff to know everything about their jobs. Instead, sales people had to know where they could acquire specific knowledge as and when they needed it. Employees, therefore, had to be integrated in a "net of knowledge", giving the 9,200 sales employees, consultants and service specialists in fifty locations in Germany quick access to a common knowledge base.

Benchmarks by management consulting firms had shown that an organization-wide network of knowledge could not only improve the quality of service, but also reduce workloads and save costs. The potential for knowledge networking at ICN was enormous. If employees could learn from past experience, if they could re-use some of the work done in creating a solution, they could offer the same solution to different customers, with only slight adaptations to suit individual requirements.

Factors to consider for the knowledge networking implementation

Setting up and sustaining such a network was far from easy. There were various factors which had to be considered first:

- How could the knowledge networking team show the sales force that they needed KN?

  According to the knowledge networking team, "You need to sell KN just as aggressively as you sell products and services". The selling of any KN initiative has to be a twofold activity: On the one hand, top management has to be persuaded of the enormous potential of Knowledge Management so that they can be won over to spread the word throughout the organization. On the other hand, the user groups have to be persuaded of the benefits.

- How would the knowledge networking team reach all 9,200 employees?

  The Knowledge Networking team suggested that this should be achieved by using an "infection" approach:

  To implement Knowledge Management in a company, you need to work with viruses. Infections must concentrate on small teams and their specific needs and requirements. These teams need to be confronted with the enormous benefits of Knowledge

Management to their particular work. As with biological viruses, the infected teams will spread the virus and infect others as the benefits materialize.

This approach was further advocated because it was perceived to create a feeling of mutual trust, since it used existing networks rather than impose new artificial ones. The virus would spread within teams as they co-operated naturally with one another, and would eventually link all the employees in a large knowledge-sharing network.

- How to define the right scope of knowledge networking?

According to the knowledge networking team, successful Knowledge Management is based on a combination of personal networks of relationships and a functioning technical infrastructure. Knowledge networking had to be an appropriate mix of "high tech" networking and "high touch" networking, reflecting the interdependence of social and technical networks necessary for effective knowledge sharing. The right "mix" of these two aspects would be crucial at ICN.

The design principles of knowledge networking

The knowledge networking team defined a set of design principles for conceptualizing the knowledge management initiative.

1. Customer orientation

A key aspect in designing the initiative was customer orientation. The initiative had to be geared to the knowledge requirements of the major user groups of knowledge networking. Sales representatives require in-depth knowledge about the competition, how to approach key customers, and how to develop solutions jointly with the customer. Service technicians, on the other hand, have a greater need for technical knowledge, for example, how complex integrated product and solution systems could best be implemented, and corresponding tips and expertise. In the initial discussions it became clear that a "one-size-fits-all" solution would be inappropriate, and that there should rather be customized solutions for special target groups. The knowledge networking team recognized that they needed to offer practical solutions that would be of real assistance for this program to be accepted.

2. Voluntary participation

The concept was to communicate the benefits of sharing knowledge for every employee and for the company. Through slogans such as "my knowledge pays off" and other appeals, employees were made to realize that any previous perceptions of gaining a career advantage by withholding knowledge from one's colleagues, no longer applied in the new competitive dynamic. People had to participate voluntarily for knowledge networking to become a reality. Prevailing mindsets had to be changed through promotion and support of knowledge networking in the form of communication campaigns, training and a bonus system.

3. Compatibility

Knowledge-networking initiatives had to be compatible with existing systems and other knowledge-management initiatives at Siemens. With this in mind, the knowledge networking team decided not to develop an independent system, but to utilize existing expertise in the areas of internal communications, training, and Intranet development. If systems were already in place, the knowledge-management processes would support these systems, instead of competing with them. Knowledge networking would therefore go hand in glove with other knowledge management initiatives. This meant, for example, that the knowledge networking team could then concentrate on the needs of ICN VD while another initiative, called ShareNet, took care of Knowledge Management between all ICN sales organizations worldwide.

4. Interdisciplinarity

As the business saying goes, "a fool with a tool is still a fool", therefore knowledge networking success requires a smart mix of behavioral changes, adaptation of business processes and the development of an appropriate IT platform. Knowledge networking had to follow an integrated, interdisciplinary approach.

5. Sustainability

The basis for lasting success in the consulting and systems-integration business would be a sustained and heightened awareness of knowledge networking. Besides the immediately visible successes, permanent change in the awareness of dealing with knowledge should be introduced. Knowledge networking should not be seen as the "flavor of the month". The quest for quick results should be accompanied by an endeavor to bring about enduring change in the corporate culture and work patterns.

With these design principles as guidelines, a special initiative to share technical tips and expertise between service technicians – Knowledge Networking Service Knowledge (KN Service Knowledge) – was launched to enable staff to respond to the increasing knowledge requirements in the installation and maintenance of individualized customer solutions.

KN Service Knowledge – networking technical tips and expertise

At present KN Service Knowledge supports the improved transfer of know-how between field service technicians (First Level Support), the Technical Assistance Center's solution specialists (Second Level Support), and the product specialists in product development (Third Level Support). It also promotes an exchange of experience within each of these levels. But before this initiative could be defined and implemented, the initial situation had to be analyzed.

The initial situation at Siemens ICN VD's technical service

The knowledge managers at ICN VD analyzed the initial situation before introducing the knowledge management system. They established the following:

- More than half the sales were achieved on the basis of the products in conjunction with know-how intensive, customer-specific solutions, and not merely the products.

- ICN had a decentralized organization and knowledge was therefore also regionally distributed.

- Solutions to customers' special needs were regularly replicated in the various regions. Past experiences were not fully utilized, or if they were, only partially and within a single region.

- The ever-shorter innovation cycles and the complexity of new solutions meant that not all the required knowledge could be acquired through formal training. Employees had to have access to knowledge as and when the needs arose.

- In the quest for solutions it was usually necessary to search many information sources in different media. Service technicians could not always verify the quality and topicality of the information.

The objectives of Knowledge Networking Service Knowledge

Using the characteristics of the initial situation as a starting point, the knowledge managers of the knowledge networking team developed KN Service Knowledge.

Oliver Holz, information manager in the Technical Assistance Center and the member of the Knowledge Management team responsible for the development of KN Service Knowledge, explained the overriding objective:

The strategic objective was to increase service quality by improving the networking of existing knowledge and collective experience within the organization. At the operative level, knowledge and experience needed to be distributed quickly and made available organization-wide, via solution maps.

Dieter Schorn, service manager at ICN VD, explained the special challenge to the knowledge managers:

Due to the large number of service staff and the many service locations, it was important that an initiative such as KN Service Knowledge was introduced quickly and comprehensively. All available opportunities had to be exploited to this end.

The target group of KN Service Knowledge comprised all service employees and their superiors who had to be persuaded to openly and actively share their knowledge with other technicians.

As the KN Service Knowledge emphasized the exchange of technical tips and expertise within the field service and between the various service levels, the process was integrated into the existing Service Information System and adapted to the specific requirements of the knowledge carriers and searchers.

Potential benefits of KN Service Knowledge

An example demonstrates the potential of KN Service Knowledge: When installing communications solutions in customer companies, the field service technicians (first-level support) are frequently confronted with existing components from manufacturers other than Siemens, which have to be integrated into the general solution. While doing this, these technicians acquire valuable know-how that could save their colleagues tedious research on similar projects – if they only knew that this know-how already existed and how to access it quickly, it would save time and money. This knowledge also represents an important tool for the remote-service call-center staff and the Second Level Support solution specialists. In addition, it provides fundamental impulses for further product developments.

The initiative also created another advantage: increased customer satisfaction due to faster service times. Knowledge Management has therefore made an important contribution to customer bonding, which means that Knowledge Networking has paved the way for sales to expand. At the end of the day, customers benefit, because they get faster, and often cheaper, solutions to their problems.

How KN Service Knowledge works

At ICN VD it was previously very difficult, or even impossible for the field service to document the knowledge acquired on site and share it with colleagues on the Intranet. This is where KN Service Knowledge comes in: The redesigned information processes permit the field staff to share their experiences easily.

Adaptation of business processes

Today, service technicians contribute their tips and expertise by phone, email or directly in a special forum in the Service Information System. This is a Lotus-Domino-based platform and, besides the application for processing service cases, also contains extensive databases with information about tried-and-tested problem solutions, examples for setting up customer solutions, hardware and software releases, software updates and other topical service information. An editorial team was set up at the Technical Assistance Center to ensure the topicality, quality and the correctness of the tips and expertise, and to prevent the duplication of entries. These experts edit and validate the contents of all information, before making it available as examples in the knowledge base of the Service Information System.

This information is also forwarded to the product development experts (Third Level Support), who adapt the set-up examples to current product specifications and ensure that the customer solutions function problem-free with future product developments and software releases. This information is also used in making decisions about new developments, as it reflects the customers' requirements and needs.

Development of a technical platform

One of the most significant barriers to a well-functioning knowledge network is a lack of time. In order to motivate staff to participate actively in Knowledge Management, it had to be possible to effect both the forwarding and accessing of information. The infrastructure at ICN VD satisfied this requirement through the availability of an existing Intranet, with almost 10,000 connected computers and a well-functioning IT system. Consequently, service technicians did not have to sacrifice much time, which made the new facility more acceptable to them.

One of the objectives was that technicians with problems should have quick and easy access to the documented knowledge, which is why the knowledge networking initiators first analyzed and then improved the query methods. As the "single point of information", the search function had to be able to scan different information rooms. It had to offer users the choice of searching for the required information on the Internet, the Intranet, in the Service Information System's databases, or in the product manuals. The search function had to be able to search sites on the Internet, Intranet, in Lotus Domino and other databases and, simultaneously, guarantee quick response times. To meet these requirements, the knowledge networking managers chose the Knowledge Query Server. This is a metacaller that takes advantage of search engines that are already in place and presents all search results in a single display.

With the new "single point of information", all staff could directly access the knowledge content of all information rooms (see Figure 14). Improved documentation and search capacities not only prevent duplicate work, but in so doing, also provide more time for personal contact, when no documented knowledge is available.

Stimulating behavioral changes – managing the human factor

The Knowledge Networking team started an information and training campaign and offered incentives to get the system up and running with sufficient input as quickly as

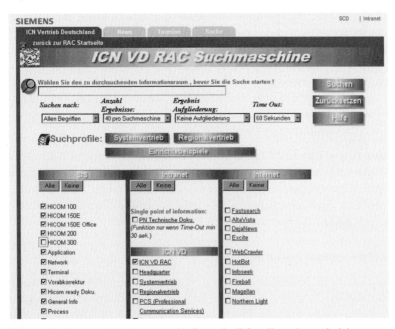

Figure 14   The new "single point of information" for all service technicians

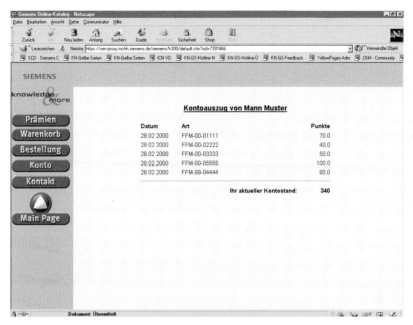

Figure 15   The "Knowledge & more" personal account statement

possible. The campaign was aimed at stimulating service staff to share their knowledge. It was also designed to overcome any reservations they may have had about using the new medium.

Motivating the service staff was the first priority when introducing KN Service Knowledge to the field. "After all, anybody planning to submit a solution is going to have to invest time", remarked Knowledge Manager Andrea Dora. The knowledge networking team developed the "knowledge & more" incentive system not only to provide motivation for the submission of tips, but also to honor the quality of the tips.

Each service employee who submitted a useful tip or trick was awarded "knowledge & more" points. Each solution submission was judged on the basis of a set of fixed criteria created by the editorial team. Only once a tip had been submitted electronically, fulfilled the minimum criteria for content, and supplied additional information about application areas and validity, would it be awarded the maximum points. When a certain number of points had been reached, the staff member could redeem the points for a bonus, ranging from personal digital assistants to travel arrangements. Bonuses were available from 300 points upwards, meaning that a single technician would have had to submit at least five solutions before being awarded the full number of points and qualifying for one of the prizes. The service technicians were able to call up their point accounts (see Figure 15) and the bonus catalogue accessed on the Intranet at any time. Apparently many found this set-up very attractive.

"After "knowledge & more" had been running for nine months, in our large-systems segment alone, more than 2,200 suggestions had been submitted. Considering that the service staff totals 2,500, this is a great result", enthused Andrea Dora.

Once the first incentive campaign had been successfully concluded, the knowledge networking team started one for the medium systems segment.

Additional motivation for the bonus system is provided at the team-management level where many regional service managers have made target agreements with their staff. The service executives thus play an important role as multipliers by motivating their teams to submit their tips and expertise. An internal corporate ranking regularly shows how many solution tips have been submitted by each service region. "This created a kind of competition and contributed to the flood of suggested solutions", reported Oliver Holz.

The service staff are trained to use the new knowledge documentation and to access the instruments. The training includes, for example, an "Intranet driving licence". In these ways, the technicians' managers repeatedly emphasize the potential that KN Service Knowledge presents.

The scope and size of the challenge that the project presented can be deduced from the above descriptions.

More about KN Service Knowledge

To support the service staff, the existing Service Information System (SIS) is being expanded to include the service employees' undocumented knowledge. To this end a process of knowledge sharing was defined. An integral part of this knowledge sharing was formed by the incentive program "knowledge & more" and the online premium catalogue.

At present service staff (of Siemens ICN VD as well as the international market) access the information site of the Remote Assistance Center (RAC), including the SIS databases, on an average of 180,000 times per month.

The incentive program "knowledge & more"

Within the framework of the KN Service Knowledge, ICN developed an incentive system, called "knowledge and more", which was available to Siemens ICN VD service employees.

The goal was to locate high quantity and quality service solution suggestions from experts in the Hicom 300 and Hicom 150 fields as part of the ICN VD business outreach, as well as to make work-related solution tips available to the employees of all sales regions.

With the temporary incentive program "knowledge & more", the solutions provided by very experienced and middle segment service employees will, through their active participation, be accessible to each and every service employee. The subject editorial staff at RAC tested every solution according to strict criteria and prepared them for inclusion in the Service Information System databanks.

The online premium catalogue

Each and every service employee can access the KN Yellow Pages through his or her password-protected entry and view his or her personal points balance in the online premium catalogue to be found there.

The "knowledge & more" online premium catalogue not only offers a view of the points balance, but also provides access to the premium offer, to the order form and to the contact site for requests for help and further questions.

When a service employee has chosen and ordered his or her desired premium from the broad offer available, the value of the premium will automatically be subtracted from his or her point balance.

The number of orders clearly indicates that the numerous solutions that were received, are of a very high quality. The requirements were exacting since the premium claim only became possible from the first 300 points onwards. However, 225 employees were able to enjoy their desired premiums.

Not all of the 3300 solutions that were received by the subject editorial staff of the RAC could meet the demanding criteria. A few of the solutions, for example, were disallowed as they were duplications.

"Trouble was taken to organize a feed-back process to the provider of each solution tip. In this way all participants received individual feed-back as to the status of their accomplishment", says Reinhard Meurs, Head Manager of Service and Information at RAC.

A solution tip also underwent the following process: The service employee sent his solution to the subject editorial staff who tested whether it completely met the needs of the required underlying principle; checked for duplication and, eventually, reworked it if required. The editorial staff also undertook the process of verification of the solution, provided the editorial input and placed the solution on the SIS where everyone can access it.

The results

Both the incentive programs presented within the framework of the KN Service Knowledge as described under "knowledge & more" above, have been completed.

The knowledge networking team has come to the following conclusion:

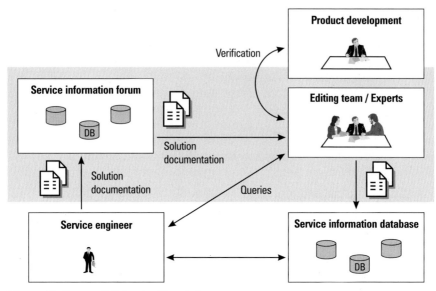

Figure 16 Process KN Service Knowledge

- The premiums offered, as well as the time limit of the programs, proved to be extremely appropriate for encouraging and directing knowledge sharing in the areas that had been specifically identified.

- Since the incentive measures were limited to a certain area of solutions as well as time-wise from the start, the raised awareness could be kept constant over the delimited time period.

- The choice of expertise and event premiums worked in favor of an identified motivational effect that went beyond the work area to affect private lives as well.

"The incentives have accomplished much", says Dieter Schorn, leader of the Remote Assistance Center. "The foundation of KN Service Knowledge has been laid in the SIS (Service Information System) databases. The system now needs to increasingly become a common place tool in the daily work of each service employee. Not everyone is as yet participating actively in knowledge sharing and exchange. Each staff member needs to motivate and encourage his fellow employees to provide solution tips and tricks and their use".

With the Service Management Award, the German Customer Service Association won the annual prize for the best customer service-oriented initiative and its repositioning.

Within Europe, the German Customer Service Association, with its 1,200 members, is the largest and most important business association for management employees engaged in customer service and service.

ously, each product sale contributes to reducing the fixed costs and, as the market expands, production leverage of knowledge assets increases greatly. Most knowledge seems to be subject to economies of scale and scope. It follows that knowledge, once created, can be deployed at low marginal costs. Still further leverage can be obtained by deploying knowledge gained in the creation process of a variety of different end products. These facts strongly indicate that a narrow focus on knowledge creation may be short sighted if a company wishes to fully exploit the law of increasing returns that characterizes the corporate environment. This is why Siemens has placed particular emphasis on a special form of knowledge transfer – *Best Practice Sharing.*

Best Practice Sharing

What does Best Practice Sharing mean? In a 1996 benchmarking project the American Productivity and Quality Center (APQC) found that eighty percent of knowledge management practitioners list Best Practice Sharing as one of their main objectives. The APQC defines best practices as "those practices that have been shown to produce superior results; selected by a systematic process, and judged as exemplary, good, or successfully demonstrated".

Best Practice Sharing constitutes an attempt to multiply existing knowledge in order to take advantage of the law of increasing returns. At Siemens, we found that the number of business options, quality improvements, cost reductions and process optimizations can be increased, thanks to the repeated use of this internal knowledge that no longer requires major investment or R&D costs. A new combination of diverse internal knowledge modules creates entirely new market opportunities. In order to exploit the law of increasing returns, our aim was to develop as many end products as possible from the present organizational knowledge base.

The critical answer that we had to find in the establishment of a best practice sharing marketplace was the definition of what is "best". We found that this definition depends on the objectives, the market conditions, our corporate culture and the culture within Groups. It is also almost impossible, or at least extremely cost-intensive, to measure the contribution of any one specific procedure or action to business success. This makes assessing what is "best" even more difficult. Difficult, but not impossible. We found that the delineation of "best" practices can only be achieved on an ongoing basis through constant negotiation and re-negotiation of what constitutes "best" on all levels of the company. Hence the rationale for establishing a marketplace as a forum for such re-negotiation.

Getting employees to share their valuable insights gained through past projects is, however, not an easy endeavor and there are many barriers that have to be overcome in order to fully exploit the law of increasing returns

Overcoming the barriers to the internal transfer of knowledge

Asking employees to share their best practices is often met with much resistance. For example, employees may not have the confidence to contribute their normal, everyday work processes and experiences as "best practices". Fundamental to the success of Best Practice Sharing is, first and foremost, the identification of the barriers impairing this. The next section describes the barriers encountered. Thereafter, successful strategies are described that were used to deal with these barriers.

Identifying the barriers to Best Practice Sharing

The very concept of something being "the best" – a superlative term implying there is nothing better – can be psychologically intimidating. It is therefore necessary for all participants in Best Practice Sharing to understand that the original contributor may label an experience a "best practice", but is more likely to be called this by a "re-user" (applying a variation of the original contribution). This is just one example of one barrier on one level. There are barriers that occur on the collective, structural and political/cultural levels as well, all of which need to be tackled. There is sometimes an overlap of these barriers and possible solutions to them, but we found that barriers to Best Practice Sharing are present on personal as well as collective levels. Furthermore, structural barriers and barriers residing in the political and cultural realm, need to be given adequate attention. For ease of exposition, the barriers encountered are stated in a tabular form.

Personal barriers

Barrier	IT solution	Network or organizational solution	Corporate solution
Don't know what others need to know.	Access for every employee.	• Involve people who are actually doing the work (Best Practice Networks) in the early stages, not only executives. • Identify major levers. • Utilize and support CoP networks.	• Best-Practice Marketplace. • Promotions. • Topic related events.
• Too much time and effort involved. • No obvious benefits or rewards.	Make it as easy as possible to use.	Incentive system, at least initially, until implicit benefits are appreciated.	• Build it formally into the working day. • Offer incentives. • Reward effort.
Lack of confidence in knowledge developed.	Computerized brainstorming – allow people to see what others have shared.	Established criteria to measure practices against.	• Give recognition to good contributions. *Success breeds success.*

Collective barriers

Barrier	IT solution	Network or organizational solution	Corporate solution
Transfer process not well organized.	Efficient tools.	• Delimit topics. • Suggest possible applications.	• IK CKM (see 6. in next section). • Facilitators edit and structure content. • Organize workshops.
In-house competition.	User-friendly system allows teams to check on their team's participation status.	Advertise section results and actively affirm participants.	• Put competitive spirit to good use and reward units/sections that participate.
• Managers are not supportive of initiative. • Poor corporate culture of promotion of best practices sharing.	Training and help for managers in use of IT system.	Include managers and key players in the planning stages of the project.	Programs like Best Practice Networks and Best Practice Marketplace help to change entrenched attitudes.

Structural barriers

Barrier	IT solution	Network or organizational solution	Corporate solution
• Keep best practices in division. • Feel that knowledge kept to oneself will help with career success.		Stimulate, actively promote, and expose staff to the benefits of knowledge sharing practices and projects.	• Network and marketplace projects. • Build knowledge sharing into criteria for evaluating performance and for promotion. • Offer incentives.
Time pressure – could be the wrong people doing the job.	Facilitate access to expertise on the network.	• Create structures for skills and expertise to be made widely known. • Match skills and expertise to tasks.	• Better selection practices. • Use (partnerships) experts from other branches and countries.
Poor IT structures.	Efficient and effective support.	Best Practice Landscape.	Sanction the necessary budgets. Find sponsors, if necessary.

Political / Cultural barriers

Barrier	IT solution	Network or organizational solution	Corporate solution
No common language.	Tool with different languages.	Structure system so that people can share in a language they feel comfortable with but which others can understand or have interpreted.	Bilingual user interface.
Competition between units.		Build into system.	Promote company-wide collaboration.
Not rewarded financially or by promotion ("Who cares?"-attitude).		• Incentive schemes. • Reward conspicuously.	• Give recognition. • Build into job requirements. • Include in promotion criteria.
Poor corporate culture – does not foster openness or build confidence. No help in dealing with conflicts.		Promotion of benefits and active involvement of managers throughout the process.	• Review policies that restrict the sharing of secret or other information. • Strong top-down promotion and endorsement.

top⁺ best practice sharing concept takes care of critical success factors

At Siemens, a best practice sharing concept was developed within the corporate-wide business-improvement program *top*⁺. Given the above barriers, implementing a best practice sharing concept had to be engineered carefully to integrate it with day-to-day practice. The *top*⁺ concept of cross-Groups Best Practice Sharing identifies six critical success factors for overcoming these barriers and for making Best Practice Sharing an integral component of the daily practice.

1. Connecting people: developing employee networks among best practice owners

We have found that employee networks are the most critical of the success factors. In employee networks (best practice networks) know-how is exchanged through the direct contact between employees. The network structure and scheduling are intended to make the process as efficient as possible for those concerned with the exchange of experience. We identified several prerequisites for this efficiency: clear delimitation of the topic, the identification of major levers for the topic, and close collaboration between the people responsible for business. It is furthermore essential to ensure that the exchange takes place between employees who are actually confronted with the topic (rather than merely being involved on a strategic or theoretical level). This is particularly important in view of intertwining Best Practice Sharing with everyday practice. Employees who show interest should learn about the particular field from bearers of know-how or best practices. This can be done in workshops where the highest possible level of interaction should occur.

For the subject-specific transfer of practices, the creation of new best practice networks is in progress, as well as the continued support of existing networks. There are already employee networks on a wide range of topics throughout Siemens, although they vary greatly regarding number of members and the radius of their actions. Unlike the networks foreseen for *top*[+] Best Practice Sharing (with a top-down structure), these networks often resulted from employee initiatives (bottom-up). One example of this is the Community of Practice Knowledge Management (CoP KM) at Siemens. In these employee networks implicit and explicit knowledge are passed on and new knowledge jointly developed.

Collaboration within Siemens is increasingly occurring between globally active teams involved in project-related work processes. Employees work together across Groups and functions to solve specific customer problems. Once a project has been completed, employees work with other colleagues on another project. This open and flexible form of networking is supported company-wide by the creation and promotion of employee networks. The common denominator of network members is the exchange of knowledge and its development regarding a business-relevant topic. The network is dissolved once the joint goal has been achieved. The promotion of employee networks therefore overlaps with, and plays a key role in, overcoming personal, collective, structural and political barriers.

2. Exchanging best practices: Best Practice Marketplace

A Best Practice Marketplace provides documented knowledge and pinpoints topic-related bearers of know-how in the company. This marketplace should make it possible for anyone either supplying or looking for practices, to find one another via project documentation. The person offering a practice describes:

- the problem,
- the problem-solving approach,
- the solution process,
- the critical success factors,
- the expense involved and
- the results.

The addresses of contacts who can be consulted in the case of queries, or personally, complete the input. People looking for a practice can present queries and publish them on the marketplace if their search is unsuccessful.

3. Relying on management commitment: engaging patrons and sponsors

At top management level we furthermore identified the commitment of managers as an essential prerequisite for the implementation of Best Practice Sharing. It became evident that sponsors needed to be found and responsibility for the topics needed to be allocated. Top management support was crucial to ensure the successful implementation of Best Practice Sharing company-wide. Management can promote Best Practice Sharing by, for example, repeatedly stressing its importance to the company. Events, communication and promotion activities were used in order to create a common context to promote the internalization of explicit knowledge, which constituted a major step toward implementing a best practice marketplace.

4. Mobilizing employees: incentives, rewards and recognition

Employee evaluation and promotion could create additional incentives for and reward the transfer and application of experience. This would ensure that the process of Best Practice Sharing is supported where there are barriers to be overcome, or preliminary work to be carried out, before the benefits become apparent. Possible levers for this are:

- formulation as an objective,
- integration in business processes, and
- specific financial and non-financial incentives.

How purposeful and successful the application of existing knowledge to new company end-products and services will be, to a large extent depends on the motivation of the employees and middle management, in particular. It is they, after all, who have to implement Best Practice Sharing on the "shop floor". Best Practice Sharing is monitored to promote its integration with the operational targets set by employees and managers. Based on Siemens' management policy, individual objectives are derived from higher Groups and departmental objectives through a dialogue between employees and supervisors. This dovetailing of objectives with the personal capabilities and development potential of individual employees, allows a high level activation of motivation. If the manager recognizes the significance of Best Practice Sharing, his insight into operational targets can be rapidly and simply passed on to other employees.

To further integrate Best Practice Sharing with day-to-day practice, Knowledge Management and Best Practice Sharing are incorporated into company-management learning programs aimed at the systematic development of leadership competence. The incorporation of knowledge transfer, as one of the tools for evaluation and promotion of employees, contributes toward the elimination of personal, collective, structural and political-cultural barriers.

5. Designing a content structure: drawing a Best-Practice Landscape

It became evident that a common language facilitates rapid and simple access to best practices. A topic structure that can be applied to all businesses (called a best practice landscape) proved to be useful here. The topic structure offers a range of experiences under which relevant experience is filed. The structure of topics is defined by the best practice landscape of the best practice marketplace. This structure represented a very powerful orientation guide and helps anyone looking for a practice to find topics of interest. The person offering a practice can file his or her contributions in the suggested topic areas and allocate keywords under which the contribution is to be found. The supplier can also decide whether to file it as a good experience or a best practice. Well-structured topics help eliminate personal, collective and structural barriers. The best practice landscape was used to support a combination of explicit, documented and new knowledge.

6. Energizing support: facilitators and "Best Practice Office"

Our experience shows that the process of exchanging experience and Best Practice Sharing needs to be supported and coordinated. So-called facilitators support, for example, the editing and structuring of experiences, and are responsible for facilitating workshops. They ensure that all necessary parameters are complied with for the exchange of experience can occur smoothly. The Knowledge Management Corporate Office (IK CKM) was entrusted with the company-wide implementation, the central co-ordination and the promotion of Best Practice Sharing. Its tasks include:

- actively establishing Best Practice Sharing across the company,
- the identification of relevant topics,
- offering support to employees for the formation of new networks,
- the co-ordinating and methodical support of existing networks,
- the co-ordinating of top^+ Best Practice Sharing with other knowledge management activities, and
- internal and external promotion.

The upkeep and further development of the Best Practice Marketplace are also team tasks. One of IK CKM's first tasks will be to establish a network of representatives (facilitators) in Groups and regions. A co-ordination and support team can help eliminate barriers at all levels.

Implementation – Toward a Best Practice Marketplace

Before turning to the pilot project that actually deals with the best practice marketplace, a preliminary project that formed the basis for the best practice marketplace is described.

Developing a basis for practice exchange: the concept of "Recruiting Network"

Collaboration within networks was identified as perhaps the most important key factor for success. The first pilot project was therefore geared towards enabling collaboration within networks, and its name, *Recruiting Network,* reflects this. The Recruiting Network was formed in 1998/99.

Such a best practice network can best be described in terms of the following attributes:

- They are networks of employees at an operational level for a specific business-relevant topic, preferably one with Siemens-wide application.
- They initially focus on Best Practice Sharing and thus provide members with an opportunity to access colleagues' of knowledge and experience with solving specific problems.
- They have a core of experts who process best practices and pass them on to employees engaged in similar tasks; best practices can also be further developed when implementation experience has been gained.
- They are supported by a facilitator.
- They function top-down or bottom-up, and should be based on a business-relevant topic.
- Roles within the network should be clearly defined and all parties should accept responsibility for their tasks.
- All those involved are prepared to share their expertise and learn from others in an unprejudiced manner.

An important question we asked ourselves was whether to implement top-down or bottom-up. Best practice networks can be initiated top-down when management recognizes that there is a particular need for a business-relevant topic and wants to stimulate knowledge sharing on this topic Siemens-wide, or bottom-up when employees recognize that their work can be made more effective through knowledge sharing and cooperation. In both cases management support is a necessary factor for success.

Clearly defined roles and responsibilities therefore form the basis of an effective network. Figure 17 illustrates how roles and responsibilities were defined in the Recruiting Network.

On top of a clear delineation of roles and responsibilities, a roadmap was built to guide the implementation of the Recruiting Network. The overall top^+ best practice-sharing concept includes a roadmap for deployment of a best practice network. Our experience has shown that each and every phase is necessary for the successful establishment of a best practice network. This is illustrated in Figure 18.

The Recruiting Network piloted the introduction of the top^+ methodology of best practice networks. The goal of the Recruiting Network project was to foster innovation and learning, and to exploit the law of increasing returns at Siemens by identifying and disseminating best practices for recruiting highly qualified employees. The Recruiting Network acted as a multiplicator in the implementation of the best practice marketplace at Siemens.

Mobilizing best practices in a marketplace

User-friendly communication media help to facilitate and speed up the forming of contacts and the exchange between employees gathering at this marketplace. The idea behind a virtual marketplace, based on this type of technology, is that it enables every employee to exchange experiences with other employees from his or her workstation.

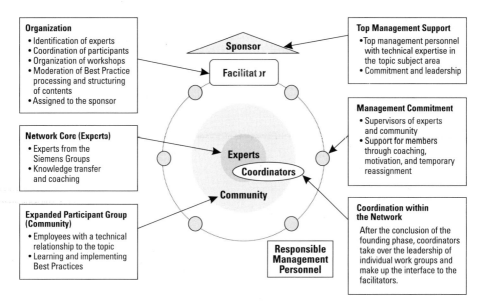

Figure 17 Defining roles and responsibilities in a recruiting network

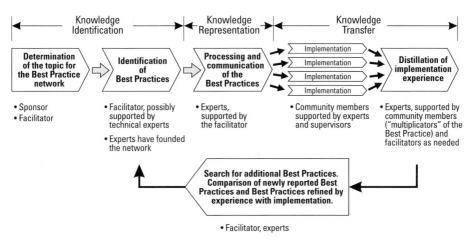

Figure 18 A roadmap for Best Practice Networks

This interaction and "trading of best practices" automatically helped to identify "best" practices.

The Siemens Medical Engineering Group (Med) launched a pilot project on the *best practice marketplace* within the scope of the *top⁺* Best Practice Sharing project. The purpose was to enable all Siemens employees to exchange experiences here, in order to find quicker, more cost-effective, or higher-quality solutions geared to improving our business.

In the preparatory phase of the best practice marketplace, the regular activities that could be influenced by the introduction of a marketplace, were identified. Interfaces and contacts were defined, management commitment was ensured and the core team was named. The following objectives were set for the introduction of the marketplace:

- Use by pilot participants
- Proof of successful implementation of best practices
- Acceptance of the tools at a quantitative (frequency of use) and qualitative level (e.g. user-friendliness, value)
- Intranet-based database system
- Fulfilling of development requirements (simple, intuitive design; self-explanatory functions – as far as possible – for those with some intranet experience, and "intelligent" search strategies to ensure rapid location of best practices)
- Fulfilling of operational requirements (trouble-free operations ensured; high access speed; potential expansion of functionality and data quantities processed ensured)
- Preparation of a Siemens-wide solution
- Validation of the best practice marketplace concept for the Siemens-wide solution
- Refinement of requirements
- Development of a pilot solution as a basis for the Siemens-wide solution.

Six divisions were defined as pilot participants within Siemens Med. The pilot project satisfied four major functions:

- The installation of an Intranet-based database system
- The mobilization of employees for the exchange of experience
- The collecting of business-relevant best practices.
- The communication of the project.

Installation of an Intranet-based database system

An external firm was commissioned to program the database system. During a 5-day workshop, the project team and the external vendors refined the required functionality of the pilot system. To start with, the requirements for an "optimal" overall system were described and these were then reduced to the functions needed for pilot operations, taking due account of cost and time limits. During the development of the marketplace tool, the individual modules were tested on a test server. The pilot system was installed on the Siemens Intranet in October 1999 and since then all employees with access to the Intranet have been able to use it.

Mobilization of employees for the exchange of experience

Experience has shown that the viability of a technical system, e.g. a database, is significantly influenced not only by the functionality of the system itself, but also by the way people react and by their barriers and fears. To identify these parameters for the marketplace, and find starting points for possible levers to eliminate barriers, employees from the pilot divisions should be able to incorporate their ideas into the project on behalf of future users. It was decided that each department would participate in a half-day workshop in order to have sufficient opportunity and time to identify the relevant barriers and thus establish a basic atmosphere of trust within the group.

The aim of the workshops was to achieve the following objectives:

- To inform employees regarding the objective and status of the pilot project
- To obtain employee requirements with regard to marketplace topics and functions
- To obtain topics for which employees could provide personal experience
- To take cognizance of and to prioritize the overcoming of barriers that impede the utilization of the marketplace
- To search for levers which could help overcome barriers
- To motivate employees to serve as multiplicators for the marketplace idea in their departments
- To secure employee commitment to support the future implementation of the marketplace at Med.

A prerequisite for carrying out the mobilization workshop was management commitment to this procedure. To this end a cascaded communication strategy was planned at management level.

As depicted in Figure 19, a cascaded communication strategy is a systematic method of approaching the distribution of communication of critical information within a large, multilevel organization. The "cascade" begins at the top of the organization and works

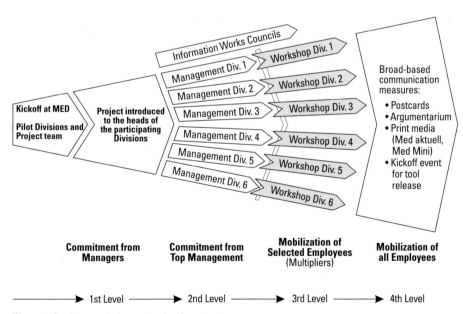

Figure 19 A cascaded communication strategy

its way down, effectively reaching each manager and employee with messages appropriate for their expected or desired reaction and contribution.

Collection of business-relevant best practices

When introducing the marketplace, it is important that the users find an adequate number of different topics when getting started. A certain amount of persistence is required for the collection of best practices. Potential contributors were called and motivated once their written requests had been received. When we were interviewing employees for interesting topics, it became evident that larger quantities of best practices can only be generated if intensive support is provided over a longer period of time.

For the pilot study at Siemens Med, a survey of the supply of and demand for best practices was carried out Siemens-wide. Contributors were requested to describe their best practices and edit them for the marketplace. It took them an hour or two to complete the profile. Telephone coaching and correction loops improved the quality significantly and at the start of the pilot project 100 experiences were entered in the marketplace.

Communication

A very important component of the overall implementation was appropriate communication of the project across all levels and functions in the organization. The pilot project was preceded by a communications campaign aimed at two target groups:

- Firstly, all employees, with the objective of mobilizing them, creating trust and eliminating fear.
- Secondly, managers, with the objective of ensuring their commitment and challenging them to create the framework for effective Best Practice Sharing among employees.

During the preparatory phase, the in-house Med journal *Med aktuell* kept readers abreast of the progress of the pilot project, while a page on the Intranet provided Med employees with information. Any outsider interested in the pilot project could keep in touch through reports in *Siemens Welt* and the *Qualifier.*

In preparation for the launch of the pilot operation, all managers at Med were sent a flyer to be used as an argumentation tool in discussions about the marketplace. A week before the pilot got underway, posters were hung in various buildings. The basic idea of making sure everyone "saw the light", was represented by a graphic design involving matches spelling out the slogan. All employees received a postcard reminding them when the pilot project was due to start. Each postcard also included a lottery number, with prizes to be won in the feedback action in December.

On the day the pilot operation was launched, Med apprentices emphasized the fact that they, being learners, are particularly dependent on the experience of colleagues. They visited some 5,000 employees at their workplace at diverse Med sites and handed out matchboxes on which with the Intranet address of the marketplace was printed. At an employees' meeting, a week after the start of the pilot, the chairman of the Med Group

Executive Management and the chairman of the Works Council stressed the significance of Best Practice Sharing and the pilot project. Throughout the pilot operation *Med aktuell* continued to report on progress. Regular feedback was used to improve the system continuously.

Summary and outlook

Substantial benefits can be reaped from a best practices sharing program if the law of increasing returns is exploited, and its managerial implications are understood. The most important implication of the law of increasing returns is that a best practice solution, once created, can be deployed at virtually no additional cost.

Siemens' experience with the best practice marketplace clearly demonstrated that knowledge-management systems are socio-technical systems and Best Practice Sharing can only be successful if the technical prerequisites are in place and the employees are willing to co-operate. Unless Best Practice Sharing is firmly intertwined with day-to-day practice of the employees, co-operation of employees cannot be achieved. The *top*⁺ Best Practice Sharing concept takes both aspects into account.

Corporate Knowledge Management was founded in October 1999, and was entrusted with the implementation of the Best-Practice Sharing concept. Corporate Knowledge Management is now in the process of implementing the global Best-Practice Sharing concept for Siemens, including derivations of it for Groups, regions and central functions. In addition CKM is establishing Best-Practice Networks for business relevant topics; providing a support center in collaboration with service providers; communicating Best-Practice Sharing internally and externally. This team is now recognized as the single point of contact for all Best-Practice Sharing topics and co-ordinates cross-Group collaboration. Steps aimed at changing company culture (e.g. agreement on objectives, incentives and process integration) are also being implemented.

The role of individual Groups, regions and central functions in the Best-Practice Sharing process is to jointly elaborate the group-specific implementations with Corporate Knowledge Management; to nominate persons responsible for the internal-group control of communication and promotion activities; to nominate persons responsible for relevant topics at the operating level to participate in networks; to mobilize employees to actively participate in the process; to find sponsors and to ensure that top-management are committed.

In these ways a solid basis has been established for the ongoing effective Best-Practice Sharing at Siemens.

Key propositions

In the course of the case study, the following key propositions for successful Best Practice Sharing could be made.

1. **Intertwining Best Practice Sharing with day to day practice to identify and share "best" practices.**

   The integration of the concept with employees' day-to-day practice is the major barrier to the successful implementation of best practice sharing concepts. Unless the best practice sharing concept is intertwined with the way employees at Siemens are doing their business, it would be a mere appendix, and unlikely to reap the full benefits of the law of increasing returns. The approach chosen to achieve this at Siemens was the establishment of a marketplace where, through negotiation of interacting parties, the definition of what constitutes "best" practices can be negotiated and re-negotiated continuously.

2. **Connecting people, rather than data.**

   In employee networks, know-how is exchanged through direct contact between employees. The network structure and scheduling make the process as efficient as possible for those concerned with the exchange of experience. Prerequisites for this efficiency are a clear delimitation of the topic, the identification of major levers for the topic and close collaboration between the people responsible for business. In addition, it is essential to ensure that the exchange takes place between employees who are actually confronted with the topic (rather than merely being involved on a strategic or theoretical level). Employees who show an interest should learn about the respective field from bearers of know-how, or best practices at workshops with as high a level of interaction as possible.

3. **The commitment of managers to the implementation of Best Practice Sharing**

   is a prerequisite for successful implementation. Management can promote Best Practice Sharing by repeatedly stressing its importance for the company. Top management involvement should therefore be part and parcel of a best practice-sharing concept. Members of the Corporate Executive Committee should accept sponsorships for specific topics. The sponsors need to support existing employee networks, encourage the creation of new networks and represent their topic areas within the company.

4. **Employee evaluation and promotion**

   are instrumental in creating incentives to reward the transfer and application of experience. It must be appreciated that the process of transferring best practices is not necessarily intrinsically attractive to the employee. Appropriate employee evaluation and promotion therefore ensure that the process of Best Practice Sharing is supported where there are barriers to be overcome, or preliminary work to be carried out before the benefits become apparent. This means that incentive systems that reward collaboration over individual performance need to be established.

5. **An "Office of Best Practice"**

   like Corporate Knowledge Management at Siemens supports all activities for Best Practice Sharing and plays an active role in the implementation of the concept.

Discussion questions

1. Describe the logic of the law of increasing returns and its implications for Best Practice Sharing.

2. To what extent is Best Practice Sharing, a special form of knowledge transfer, different from knowledge creation? Do both lead to the same outcomes? Substantiate your answer.

3. Critically discuss the barriers to Best Practice Sharing and analyze in what ways these differ from the barriers to knowledge creation.

4. Critically analyze Siemens' sophistication in overcoming the barriers to Best Practice Sharing. Do you think that the critical success factors identified address all the barriers?

5. Why was it useful to apply the concept of a marketplace in the context of Best Practice Sharing?

6. "Employee networks are the foundation of Best Practice Sharing." Discuss this statement critically through the lens of Communities of Practice.

III   Communities of Practice

The power of communities:
How to build Knowledge Management on a corporate level using a bottom-up approach

Ellen Enkel, Peter Heinold, Josef Hofer-Alfeis &Yvonne Wicki

Abstract

We frequently say that people are an organization's most important resource, yet we seldom understand this truism in terms of the communities through which individuals develop and share the capacity to create and use knowledge. Even when people work for large organizations, they learn through their participation in smaller communities made up of people with whom they interact on a regular basis. At Siemens, a group of committed knowledge management staff regularly discussed the latest developments in the knowledge management field and helped to find solutions for each other's most difficult challenges. This group formed a Community of Practice for Knowledge Management, sharing a common background, practice and identity, and engaging in a common enterprise. In 1998, this Community of Practice requested central support from Siemens. This "bottom-up movement" led to the creation of a new corporate office for Knowledge Management, the Corporate Knowledge Management Office (CKM).

In this case study the phenomenon of Communities of Practice is examined as a driving force for effective Knowledge Management in a company. The challenges facing its set-up and successful implementation are explored. As a forum for sharing knowledge and a knowledge community, and as a testing laboratory for integral knowledge management systems, its strengths and critical aspects are discussed. Thereafter, the factors that contributed to its successful creation are examined. In particular, the question of how this Community of Practice in Siemens a non-centrally organized and heterogeneous company was able to develop and attract sufficient attention to bring into being a new corporate office is examined together with its aim to co-ordinate and support the knowledge management activities of Siemens.

Introduction

"I expect he will soon be promoted to a job in the corporate department," says head of department, Laura Schnell, to her colleague Thomas Falter. "I shall certainly miss his

knowledge and experience when we tackle new projects. Deadlines are going to be more horrific than usual and I am not sure whether I shall be able to keep to the deadlines in these circumstances. But I'm delighted for him that he's been given this step up the ladder; he really does do excellent work."

"Yes, one should actually expect capable employees to aim for higher objectives and then take their knowledge and considerable experience with them," replies Thomas. "There's a similar problem with retirees. Unfortunately, we still do not have a procedure that would enable us to retain their profound knowledge or to transfer it to a successor."

"We should think of developing a concept that will take these kind of situations into account. Then we could offer different approaches to the problem. They could include different methods and techniques for retaining specialist employees, building a feasible co-operative relationship over the medium term, systematically transferring knowledge, training successors and documenting core knowledge that will have value over the long term," Laura suggests enthusiastically.

Thomas reminds her: "There is a way of developing a solution that includes different areas of expertise and provides various examples of solutions. If anything can be described as a "living knowledge base" of our company, then it is our Community of Practice Knowledge Management."

Laura is deep in thought. It may be possible to keep her employee, or at least secure his knowledge for her department. "If I were you, Thomas, I'd take immediate action!"

He turns away to send an urgent e-mail request to the Community via its mailing list.

From a Community initiative to a central corporate office

The phenomenon of the bottom-up approach

What is referred to here as a phenomenon of the bottom-up approach is a Community of Practice initiative that led to the establishment of a central office for the interlinking of knowledge management activities at Siemens.

The phenomenon started in 1998 with a request for central support by the previously informal Community of Practice Knowledge Management. This, as the name implies, is a knowledge community concerned with Knowledge Management. A Community of Practice is a group of people who are linked together by a common ability or a shared interest, and consequently possess common practical experience, specialist information and intuitive knowledge. They share information, experience and insights and are supported by various tools.

The Community of Practice Knowledge Management (CoP KM) at Siemens is a company-wide community of people who are active in the field of knowledge management. This Community was formed in 1997, starting off with 15 members. Over the years,

Siemens initiated various knowledge management activities and projects all over the world. People who gained experience through these activities and met (more or less accidentally) began to exchange their experience and their knowledge of handling knowledge. If a problem occurred, they would get in touch with each other. Informally, they began telling one another stories about their successes or failures in the handling of knowledge until, finally, they formed the CoP KM. In the development cycle of the Community, regular face-to-face meetings occupied an especially important position.

The like-minded participants used their first meeting, which took place in January 1997, to exchange experiences and to commit themselves to continuing their mutual co-operation. The basic Community was thus established and attracted more and more specialists and interested staff. The Community developed according to the principle of open knowledge communities: it had no particular mandate and participation was voluntary and open to everyone who was or wanted to be active in this area – from factory students to top managers.

Other important stages in the history of its development were the meetings in July and December 1997, when a common understanding of knowledge and Knowledge Management (tasks, applications, instruments and processes) was developed, issues were identified and areas of responsibility allocated. In addition to this, the December meeting focused on organizational integration. The size of the Community, and the volume of work it did (over a hundred members from all parts and different levels of the company attended this meeting), required support, co-ordination and the provision of resources.

The Community grew rapidly as a result of an ever-increasing interest in knowledge topics and the perception of the Siemens staff and management of the enhanced importance of knowledge. Its size made its continued existence as a self-organized community of employees concerned with knowledge-related topics, impossible. At the same time, the Community wanted to involve more staff in actively contributing towards the transfer of knowledge across all hierarchical and Group levels. Such an undertaking, however, was no longer possible without central co-ordination and organization. In 1998, therefore, the Community officially requested support, triggering an astonishing process – the phenomenon of the bottom-up approach.

As there was no central office coordinating and organizing cross-Group knowledge management activities at the time, the CoP KM initiated the establishment of a taskforce at corporate level. The core of this taskforce consisted of various members of the Community. These included Peter Heinold from the Corporate Planning and Development Department, Dr Josef Hofer-Alfeis from the Corporate Technology Department and Peter Vieser from the Corporate Human Resources Department. They were supported by a high-ranking steering committee made up of managers from Siemens' corporate departments, and headed by Professor Pribilla, a member of the Corporate Executive Board. This composition guaranteed – like the Community itself – an interdisciplinary form of co-operation. The Community of Practice had become formalized.

What are Communities of Practice?

Within just one year, the taskforce succeeded in preparing its agenda – the expansion and support of the Community of Practice's knowledge-related activities through a central department – and in communicating the Community's significance for the company to its management. But what exactly is this phenomenon (Communities of Practice) that has, in recent years, come to be regarded highly both in science and in company practice?

Investigations were made into how people exchange information and communicate with one another in everyday company life. It was established this is exactly the way in which Communities of Practice are formed: by a group of people sharing their practical experience, specialist skills and intuitive knowledge about a common interest, with each group developing its own social and cognitive repertoire governing its actions and interpretations. The process of knowledge-exchange takes place on an informal basis and the members of such a Community develop a single identity – as well as shared values and knowledge – by solving common problems, becoming involved in their mutual work and sharing their everyday concerns. They form a Community of Practice.

From a task force to a central corporate office for Knowledge Management

The meetings of the Community played an important role in the further development of the knowledge management initiative at Siemens. In September 1998, for example, the participants discussed whether knowledge and the handling of knowledge, as developed among the members of the Community, could also become an independent business. The question whether, in the light of the professionalism and efficiency attained by the Community, it was possible and desirable to provide these skills to other companies as well, was thoroughly examined.

In February 1999, the meeting concentrated on organizational integration. A discussion was held on what the planned organizational structure and the tasks of the future CKM organization would look like and what requirements would have to be met. The CKM Council was also formed at this meeting by Peter Heinold. The aim of this council was to identify official representatives of the business units to serve in the Community. These representatives would build bridges between the corporate departments and the Groups and between corporate projects and Group projects.

The task force was able to provide such a convincing picture of the potential inherent in the Community of Practice, and the value of a central organizational unit, that the Corporate Executive Board decided, in July 1999, to create a new corporate office named Corporate Knowledge Management (CKM). The CKM office was to be assigned to the Corporate Information and Communication Structures (IK) Office, headed by Chittur Ramakrishnan, the Chief Information Officer, and take on these challenges with its own staff under the leadership of Günther Klementz.

CKM got down to work on 1st October 1999. Heinold, Dr. Hofer-Alfeis and Vieser, who had initiated the transformation from an informal knowledge Community to a formal Community of Practice and the establishment of a new corporate organization, had managed to achieve this in only one year! The integration of CKM into IK led only a few months later to a new name for IK. As of 1st April 2000 Chittur Ramakrishnan changed the name of IK, previously German for Informations- und Kommunikations-strukturen, to now English Information and Knowledge Management!

Figure 20 illustrates the astonishing development of the CKM organization at Siemens as a result of the activities of CoP KM.

Due to its central organization, the CoP KM continued to develop. The meetings in October 1999 and May 2000 were particularly significant. The October meeting concentrated on globalization. The Community had decided to initiate contact with experts and networks all over the world and organized an initial international meeting in America (meetings in Asia and Australia were to follow). At these meetings, knowledge management activities in the relevant countries were to be integrated and the Community's German projects presented.

In May 2000, an international meeting was held in Munich for the purpose of self-reflection and to deal with open questions. Over 200 managers and members of the CoP KM from many different countries met over a two- day period. Those of the more than 350 fifty members of the CoP KM throughout the world who were unable to attend personally, were able to follow the conference live on the Intranet. A Community of Practice assessment, carried out just prior to the meeting, which had suggested measures for improving Community performance and increasing its benefits, triggered a discussion about the future development of the Community of Practice. Since then regular conferences every 6 months were organized by CKM to further develop the topics of KM for Siemens and to foster the networking of the community members. In December 2001 the CoP KM celebrated already its 10th meeting and dispite tight budgets and a conference fee more than 80 members attended two days. Further conferences are planned.

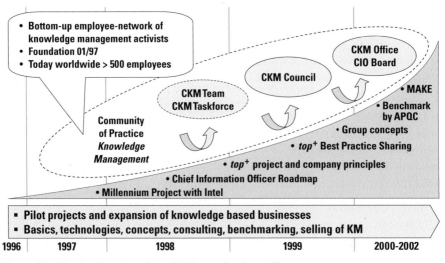

Figure 20   The development of the CKM organization at Siemens

Tasks of the CKM office

The tasks of the CKM office are by no means restricted to supporting the Community of Practice Knowledge Management. They also include many additional activities surrounding the knowledge management topic. With more than 100 projects currently running throughout the company, the subject has gained enormous impetus and the trend is increasing. The need for constancy, transparency and a joint approach, in other words the need for more than just a loose association of people who are active in the knowledge management field, has therefore grown as well. Thanks to the work of CKM, Knowledge Management at corporate level has acquired a formal platform, with its own special mandate and resources for the very first time.

However, the CKM itself is a Community of Practice as well. In order to justify its role, the CKM office works in a company-wide knowledge management network and acts both as a pivot and a catalyst. Its main task is helping business owners meet their daily challenges by applying knowledge management solutions, for example, with regard to processes of innovation, customer relations, increasing quality and E-business. To this end, CKM offers processes for developing knowledge management strategies and integrating them into business strategies. It also provides information and develops tools for support purposes. Another important task is to increase the motivation of those who share their knowledge, in order to achieve a common understanding of the advantages of systematically sharing and creating knowledge for the individual and the company alike, both now and in future.

The fact that the CKM has built the "Community of the Communities" shows, interestingly enough, that this concept is replicable. By having evolved from an informal to a formal Community of Practice, increased in size and obtained support from different sides, the idea and values of the initial Community of Practice have been reproduced and the experiences and learning handed over. Figure 21 gives an overview of the CKM office's activities.

Knowledge Management activities at Siemens

Today, five years after the first initiative by the CoP KM, the CKM office has succeeded in establishing itself as a forerunner, adviser and promoter of Siemens' Knowledge Management. Thanks to the commitment it has shown and the success of its work, it has gained acceptance across the boundaries of in-company Groups, countries and hierarchies. It can therefore be interpreted as pioneering Siemens' re-alignment as a knowledge-based enterprise.

On the basis of this fundamentally democratic approach, many other diverse initiatives and projects supporting the coordinated expansion of Knowledge Management at Siemens, have been started in recent years.

Figure 22 shows the wide variety of the projects and activities that CKM promotes and supports. As far as CKM is concerned, whether these projects are carried out under its name or not, is unimportant; CKM is far more interested in providing competent support for the knowledge-based activities of the individual Groups and departments.

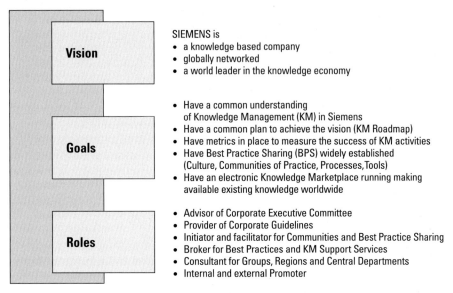

Figure 21 Tasks and responsibilities of the CKM Office

An international comparison of these knowledge management activities shows just how successful Siemens' Knowledge Management has become since it came into being. Teleos, for example, a leading knowledge management research institute, regularly draws up a list of the most highly regarded "knowledge enterprises" in the world. In the survey of the year 2001, in which the opinions of 300 knowledge practitioners in management functions were recorded, Siemens is the first German company to appear among the top twenty companies for the forth time (www.knowledgebusiness.com).

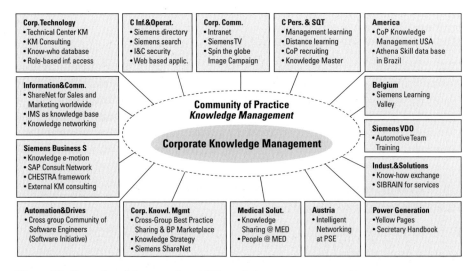

Figure 22 Examples of the more than 100 knowledge management projects

- **Most Admired Knowledge Enterprise (MAKE) by Teleos, Great Britain:**
  Siemens selected 1998 - 2002 as best German KM company by Fortune 500 CEOs/CFOs
  and 300 KM experts worldwide:
  - 1998 - 2001: ranked number 7 in 2001 out of over 100 nominated companies worldwide
  - 2001/2: ranked number 5 out of 91 nominated companies in Europe
- **APQC, USA: International Consortium Benchmarks on KM:**
  Siemens selected 1999 - 2002 as Best Practice Partner
- **MACILS, Germany:**
  Siemens selected 1999 – 2001 as KM Benchmark for German companies
- **KVD, Germany:**
  Service Management Award 2001 for Knowledge Networking in Telco-services
- **Numerous international conferences:**
  Siemens invited for key notes, speeches, workshops
- **Industriearbeitskreis Wissensmanagement in der Praxis:**
  Siemens as founder for 50 companies on KM Best Practices, www.wimip.de.

Figure 23 External recognition of Knowledge Management at Siemens

Further confirmation of just how successful Siemens' Knowledge Management is, can be found in the February 2000 report of the recognized American Productivity & Quality Center (APQC). APQC (www.apqc.org) selected Siemens from among 80 candidates as one of the 5 global organizations that can be regarded as models for the successful implementation of Knowledge Management. APQC is a source recognized worldwide for improving the processes and performance of organizations – regardless of their size and the sector of industry to which they belong.

Siemens' method of organizing corporate Knowledge Management enables any knowledge-intensive business to implement its own knowledge management activities more rapidly without having to rebuild its know-how repeatedly. It also shows companies how to recycle their intellectual capital and thus facilitates the swifter realization of innovations. The CoP KM has been especially singled out in this regard. It is regarded as an effective and powerful way of generating, sharing and transferring knowledge. In the opinion of APQC, the experience gained by the Community of Practice Knowledge Management has become a core accumulation of intellectual capital.

How much Knowledge Management at Siemens is recognized from external point of view, can be seen from Figure 23.

The heart of Knowledge Management at Siemens – the Community of Practice

A glance at Teleos' rankings list, reveals something else of considerable importance, namely that no company today can afford not to look for ways to make the best use of its knowledge. In view of the processes of rampant change in the global markets, resulting in a completely new value being ascribed to knowledge and its intelligent networking, this is no surprise.

The importance that Siemens attaches to the handling of knowledge is indicated in its mission statement, which includes the objective of creating a company-wide network of knowledge. And it is exactly here that the difficulty – or challenge – lies. According to a recent study, only 20 to 40 percent of the knowledge in companies is actually being used at present – despite all efforts to date. Even Siemens, with its unequalled range of know-how, still shows enormous potential in the networking of different sources of knowledge. This potential must not be left unexploited. Providing customized turnkey solutions is often not enough; they have to be provided quickly in order to beat the competition.

Knowledge Management is aimed at promoting knowledge processes, such as the "localization" and "recording", "dissemination" and "accumulation" of knowledge, and doing so purposefully by means of various activities. According to Probst and his team members of the Geneva Knowledge Group, Knowledge Management can be defined as "an improvement of organizational capabilities on all levels of the organization through better handling of knowledge as a resource". A study by the Gartner Group has shown that most large companies have started to implement knowledge management schemes, or are in the process of developing them.

A successful Knowledge Management method is the concept of Communities of Practice. The simple fact that the changes initiated by the Community of Practice Knowledge Management as a bottom-up movement at Siemens resulted in such considerable success, is a clear demonstration of the importance and value such Communities of Practice can have for a company.

First results…

After three days, Laura Schnell and Thomas Falter take stock. How successful was their e-mail enquiry to the Community? Will there be only well-meaning advice in answer to their query, or will they receive really helpful information? Perhaps the finding of a solution to the problem of "departing experts" doesn't interest anybody!

The two staff members are fortunate – the results of their efforts have been extremely successful and even exceed their expectations. After only 3 days, they have found 9 additional comrades-in-arms from various Siemens areas and departments who are confronted with similar problems. They come from the development, personnel, training and knowledge management sections of the company. An ideal combination!

Apart from these people who are interested in active co-operation, many other Community members have replied as well, including several interested parties from a regional knowledge management Community currently getting established, i.e. the Community of Practice Knowledge Management in North America.

"Fantastic!" Laura Schnell proclaims jubilantly. "We have brought together the necessary fields of expertise needed to develop a comprehensive concept, and we can also test its applicability to other countries immediately!"

Thomas is equally enthusiastic. "I never imagined that we would stir up such a hornets' nest with our problem of "departing experts"! Now we only have to create an official sub-Community for this issue. We will have to obtain the support of the CKM office. It has many years of experience and a great deal of expertise in this area."

Laura and Thomas approach the CKM office, who not only supports them in creating the sub-Community for the "departing experts" problem, but also takes up BMW's recent offer to collaborate on a similar topic. After a few telephone calls, BMW is ready to work with the Community to learn from Siemens' successful Community concept. This seems an ideal opportunity for a classic win-win situation.

Three members of the new "departing experts" Community are invited to a project meeting currently being planned with a well-known research institute that is developing a system for BMW to record expert knowledge by means of interview techniques, terminological systems and hypertext documentation.

The power of Communities of Practice

The phenomenon of Communities of Practice first appeared in the scientific literature at the beginning of the '90s. A study, which investigated service technicians at Xerox, revealed that there were considerable differences between the formal description of their work and the actual procedure followed when they had to remedy defects in copying machines. Whenever the service employees were confronted with a specific problem for which their manual provided no solution, they looked for informal sources of knowledge. In fact, even before consulting the manual, they would ask their colleagues for help and tell one another stories of problems that had already been solved. The result was a Community of Practice in which the employees exchanged information with each other on a regular basis. The service technicians developed specific communication structures, customs and a shared identity. On the basis of this study, this phenomenon had been conceptualized into and called a Community of Practice.

Communities of Practice are everywhere – we all belong to a number of them, whether at work or in private life. We may lead a group of experts who specialize in human resource strategies, or we may just stay in touch to keep informed about developments in the field. Whatever form our participation takes, most of us are familiar with the experience of belonging to a Community of Practice. In organizations they can be hidden and are often difficult to detect. Since their members come and go Communities of Practice are also continually fluctuating. These circumstances present organizations with great difficulty in identifying such Communities of Practice, determining their borders and identifying methods for supporting them. However, it was possible to identify and support the Siemens Community of Practice, because it went public when it requested central support and, due to its bottom-up movement, brought a new central organizational unit into being.

Knowledge exchange in Communities –
How Siemens supports the transfer of knowledge

A Community of Practice represents a common body of knowledge. They therefore exchange knowledge efficiently. The members achieve this efficiency by exchanging anecdotes, for example, about specific solutions to problems. The internalized knowledge is not written down, but is exchanged directly. The exchange of knowledge in Communities can take place explicitly or implicitly, thus the distinction between explicit and implicit knowledge is crucial. The articulated, or codified form of knowledge is explicitly represented in physical or material objects. These can be patents, manuals etc. Explicit exchange involves people's own language or vocabulary, codified procedures, relevant documents and so on. Implicit knowledge is difficult to formulate and therefore difficult to communicate to others. In order to ride a bicycle, we need to know how to keep our balance. We do not think about whether we should steer to the left or right to avoid falling off and, if we were asked, we would be unable to articulate what exact knowledge is needed.

This implicit knowledge is rooted in our everyday behavior and is always connected to a specific context – a specific technology, a profession or a community. Our "know-how" – the practical skills or expertise that allow us to work efficiently and effectively and not always having to think about the detailed ways of solving a problem, but simply doing it – has its origin in our implicit knowledge. However, implicit knowledge also has a cognitive dimension; we possess it in the form of embedded mental models, beliefs and perspectives, so that we regard it as reliable and unquestionable. This knowledge represents the invisible relationships, or glue, which keeps Communities of Practice together. The Community of Practice is thus efficient in transferring and dealing with implicit knowledge which is tied to the social context, embodied in language and behavior, and is, therefore, difficult for a company to access and make useable.

At Siemens, the esteem in which the Community's work is held, and the support provided by a suitable infrastructure, keeps the Community alive to generate further activities. An example of such support is provided by the Community conferences, organized by CKM, which take place at regular intervals and enable social contact among the members, thus also promoting the exchange of knowledge.

Due to their informal nature and their constant development and self-organization, it is difficult to define the borders of Communities of Practice within the traditional company gridlines. Communities of Practice may go beyond the borders of a company by involving members of other companies who are working on similar problems. The real borders of Communities of Practice are defined by the shared knowledge. They provide the group and individuals with a strong feeling of identity and are thus not based on knowledge alone, but also on emotions and a feeling of responsibility for and commitment to the company.

The Community of Practice Knowledge Management has developed from an informal network of knowledge management practitioners or interested people, into an association that stretches beyond the borders of organizational units and countries. But what does it mean to support or maintain such Communities?

menting the measures and proposals of the Community of Practice and by achieving success in the process. The CoP KM, for example, achieved success through their request for support.

The "departing experts" Community

In October 2000, one month after the first Community activities on the topic of "departing experts", the sub-Community (in which two BMW employees took an active part) presented its astonishing solution to the Community meeting.

"I find your approach extremely interesting. How did you succeed in coming up with such a far-reaching and versatile solution?" a colleague asks Laura Schnell.

"It wasn't too difficult," she replies smiling, "but I have to tell you that without the help of the community and the facilitator from the CKM office, we would never had done it so quickly. They supported us whole-heartedly, for example, with guidance and the organization of our sessions. Their purposeful and systematic way of going about things almost automatically led us to constructive results."

"I believe that the Community as an organizational form also contributed to this result," counters Thomas Falter. "From the very beginning I felt as if I had been adopted into a knowledge Community that had similar problems and almost the same interests. This led to a very relaxed and constructive atmosphere."

"Then we should also try this form of organization in other areas. After all, today's results speak for themselves!" replies their colleague.

Outlook

The increasing importance of Knowledge Management

By promoting the knowledge activities of a Community of Practice, even to the extent of creating new departments and fields of responsibility, Siemens is well on the way to becoming a knowledge-based company. Siemens is convinced that effective Knowledge Management will increase the value added in future. To this end the forms of organization are not only being adapted and new ones created, but also management systems are being redesigned. Knowledge Management is of outstanding importance to the company, since it generates long-term competitive advantages, which are particularly important in the light of Siemens' increasing involvement in the services and consulting business and the transformation into an e-business company.

The trend towards services and consulting business has led to an increasing demand for knowledge. New knowledge-based business is therefore gaining importance. Knowledge business includes all kinds of marketable knowledge such as services based on specialist knowledge (consulting/engineering, servicing/training and management/financing), procedural models and access to the company's knowledge networks (competence networks and knowledge management processes). In more precise terms, this means access to process models and experience based on the company's own operational knowledge, e.g. business-process models and project experience, or experience and methods arising from business transformation. Access is provided to Siemens' own areas of competence and knowledge management systems, for example, by giving advice on the whole internal range of technology and by allowing access to the Best-Practice Exchange and to the Communities of Practice.

Knowledge Management will soon become a business in its own right. This means, in figurative terms, that we are not only selling the cake, but also the recipe for making it. This also means that the work of the CKM office involves the following tasks: The development of a knowledge sharing culture, definition of a knowledge strategy and framework (incl. standards and guidelines) and alignment of existing KM activities, development and rollout of a CKM roadmap, initiation and support of Communities of Practice (CoP), introduction and promotion of best practice sharing across groups, regions and central departments, provision of a Best Practice/Knowledge Marketplace for company-wide knowledge exchange, definition and introduction of reference architectures for Knowledge and Information Management systems and working enviroments, definition and introduction of qualified services for Knowledge and Information Management activities.

Additional activities of the CKM office

Many projects that go beyond the boundaries of individual departments and Groups, have already originated from the Community initiative.

CKM is assuming the responsibility for Siemens-wide implementation of best-practice sharing. This includes, above all, the expansion of cross-Group staff networks in the context of top^+ best-practice sharing, the establishment of an electronic best-practice marketplace and the ShareNet project. In addition to these projects (also presented in this case book) the establishment of a complete Knowledge Sharing system and Knowledge Communities at Medical Solutions should also be mentioned, as well as Siemens Business Services activities relating to Knowledge Management (knowledge-motion), and the Siemens Learning Valley project in Belgium. New activities emerge in Brazil, India and USA. Many of these projects and activities are mentioned in this case book and are intended to demonstrate the great variety of possibilities open to a company like Siemens.

The CKM office's mission is to promote company-wide transparency of knowledge and knowledge holders, to avoid the creation of isolated knowledge management solutions, and to contribute to the development of a knowledge culture. Examples of these are the development and improvement of a knowledge management road map in association with the Groups, and the initiative for disseminating the ideas contained in the company'smission statement.

7. Using organizational forms like the Communities of Practice, and a central organizational unit, such as the CKM office, to implement Knowledge Management will assist in the transformation of a company into a knowledge-based enterprise.

Discussion Questions

1. Why is it possible to speak of the power of Communities of Practice?
2. Why can Communities of Practice be designated as the heart of a knowledge management system?
3. What is particularly characteristic of Communities and what value do they have for the company?
4. Identify the factors that determine the success or failure of Communities of Practice.
5. Can the Siemens' Community concept be applied to other companies? Why?
6. What value does a central knowledge management organization like CKM have for the development and continuation of a Community?
7. What tasks should a central knowledge management organization perform?
8. What does the future hold for a knowledge-based company? What concepts and forms of organization can help pave the way for this?

KECnetworking – Knowledge Management at Infineon Technologies AG

Michael Franz, Rainer Schmidt, Stefan Schoen & Sabine Seufert

Abstract

This case study describes how a manager and his team implemented and systematized the exchange of process-oriented knowledge within the factories of Infineon Technologies (manufacturers of silicon microchips), in five countries on three different continents, in order to solve manufacturing problems and to optimize procedures. The ambitious business goals of the company could only be realized if all ten sites were able to achieve the same level and quality of output. This would be possible, the manager felt, if valuable expertise were easily, continuously and freely accessible, as and when the problems arose, and not only sporadically, as had been the pattern in the past. To this end, they initiated the Knowledge Exchange Networking (KECnetworking) project in 1990, in cooperation with the Knowledge Management Department of Siemens AG Corporate Technology. The KECnetworking, as the project became known, is ideally suited to serve the ever-changing and highly competitive field of semiconductor technology, where strategic technological knowledge has a relatively short "half-life" and thus needs to be transferred and implemented as quickly as possible.

The case begins with a brief history of the development of the microchip, followed by an overview of the KECnetworking project. Not only is the current structure of the knowledge exchange then examined, but some of the pitfalls encountered during the development of this initiative are reviewed, as well as the steps taken to make improvements.

"The permanent, personal knowledge exchange among small expert teams with the same expertise, leads to stronger personal relationships and networks. In this way trust is built among the participants – a prerequisite for efficient best-practice sharing and for overcoming the "not invented here" syndrome. And finally: a driver is needed who will organize the continuous exchange of knowledge and who will watch over the ongoing exchange". Dr. Helmut Gunther, Infineon Technologies AG, Senior Director Memory Products Frontend Productivity Improvement

The corporate context

Infineon Technologies AG – A new name with a long tradition in semiconductors

On 1 April 1999, the Siemens semiconductor division was newly established as Infineon Technologies AG. A little less than a year later, on 13 March 2000, the company was publicly listed. During this process, the name Infineon became widely known, and the initial stock offering was oversubscribed thirty times. With the exception of Deutsche Telekom, Infineon had become the second largest initial public offering (IPO) in Germany.

The company is no newcomer. It was founded on a long tradition that goes back to 1952, when the first factory, called Siemens Halbleiter (HaF) – Siemens semiconductors – was established. During the last five years, the firm has ranked 10th overall in the ratings of the world's largest electronic microchip producers. As a multinational company, Infineon has become a globally intertwined network of development and production sites and one of the most aggressively expanding companies in the industry. The company had revenues of around four billion Euros in the 1998/99 business year and has more than 25,000 employees, of whom fifty percent are outside Germany. Infineon attracted particular attention with this year's IPO, but the firm's origins can be traced back much further.

1947-1957 – a decade of transistors

In 1947, the invention of the transistor presented an alternative to the vacuum tube. At that time the newly emerging semiconductor industry was already characterized by the same attribute that is still its hallmark today: the belief that progress lies in miniaturization. The transistor, which was much smaller than the vacuum tube, produced less heat and consumed less energy, meant that products could be built much more compactly than ever before. The transistor radio was born.

Barely five years later, Siemens founded the Siemens Halbleiterfabrik (HaF) or semiconductor factory. The following year the first germanium transistors went into production. The firm's headquarters is still situated in Munich where it moved in 1953.

A new semiconductor leads to rapid progress

In 1957, production of a new kind of component commenced using silicon, a non-metallic element that possesses important semi-conducting properties. Six years later, in 1963, Siemens started production of integrated circuits, which had replaced transistors and are crucial in the construction of computers. At the time, Siemens was continually expanding, and had established several production factories in Germany and Singapore. Today Infineon controls ten state-of-the-art production sites in five countries, including memory chip factories on three continents: Europe, Asia, and the USA.

A volatile and extremely competitive market

The semiconductor market has been marked by rapid growth, propelled mainly by the proliferation of the personal computer. During 1992-1999, the global market grew at an average rate of 10.2%, with total revenues increasing from $US 50 billion to $US 55 billion. During the same period, Siemens/Infineon's growth was twice that. The strong market was offset by equally strong volatility with the industry fluctuating between – 20% to +50% during the last ten years.

Siemens was not spared these market fluctuations, or the enormous competitive pressures. The global excess capacities of 1997/98 led to a decrease in the price of memory chips. This forced Siemens/Infineon to close a production site in the UK. The newly established factory in Dresden, which the media had previously hailed as the "Silicon Valley of eastern Germany", also came under intense pressure to perform. On the whole, this market is, arguably, a future one. Electronic intelligence will become an integral part of our lives. The industry has coined the term "ubiquitous computing" and examples of what is envisioned are "interactive cars" and the concept of "intelligent houses". Technological progress is, and has been, based on miniaturization, compactness, economy and improvement in memory chip performance. Whereas the first cellular phones weighed roughly 500 grams, the newest models weigh less than 100 grams and they are significantly cheaper.

The semiconductor industry presently differentiates between three types of chips, namely, microprocessors, logic chips and memory chips, all of which are produced by Infineon. While microprocessors and logic chips are minute computing machines, memory chips (also called DRAMs) provide the information that is necessary for computing.

Rapid technological development and the race to be first to bring a new technology to the market are especially fierce with DRAMs. Development has occurred in leaps and bounds, often cooperatively. In 1981, the 16 kbit DRAM was produced; in 1985, cooperation with Toshiba resulted in the 1 Mbit DRAM and in 1990, cooperation with IBM led to the production of the 64 Mbit DRAM. From 1992 on, cooperation has resulted in steady progress, the most recent development being the 256 Mbit DRAM, in cooperation with IBM and Toshiba.

This short history illustrates two important points about the nature of this industry. Firstly, it highlights the increasing pace of development and the rapid decrease in the "half-life" of knowledge. Secondly, it emphasizes the importance of cooperation and joint ventures with competing firms, as with IBM and Toshiba.

Partners in development

Despite the intense competitive forces and pressure to bring new products to the market, Siemens recognized the importance of cooperation and strategic alliances at an early stage. Siemens/Infineon not only cooperates in the area of technology development with IBM, Toshiba, and Motorola, but also counts NEC, Nokia, and SONY among its strategic partners in application development.

The company's primary motivations for cooperation and network formation are:

- The sharing of risk and development costs
- The reduction in product "time-to-market"

This takes on special significance in light of the growing complexity of the circuitry production process of integrated circuits.

ICs and their production become more complex

The basic raw material required for the manufacture of integrated circuits (ICs) is a silicon wafer. The wafers are manufactured from raw silicon which occurs naturally as silica in sand and quartz. This silicon is then purified and cultivated in a single crystal form. These crystals are then machined and sliced into thin wafers. ICs are manufactured by depositing layers of various chemicals and gases on the wafer, etching away the unwanted areas, and then implanting various isotopes in certain areas to achieve the desired electrical parameters. In this way, thousands of transistors are manufactured on a few square millimeters of silicon. These transistors are then interconnected to form logic gates (switches), which are the backbone of any digital circuit. The minimum dimensions to which these ICs are manufactured are as low as 0,18 µm (i.e. one millionth of a meter), or one thousand times thinner than a human hair. ICs have already been manufactured on an experimental basis to a minimum geometry of 0,13 µm.

Given the very small dimensions of ICs, the manufacturing process must take place in an extremely clean environment as any dust, or other particles, will contaminate the IC. Such clean-room factories cost several billion US dollars to set up.

The chips are produced on so-called wafers, a silicon slice on which, depending on the size, two thousand or more chips can be placed. As long as the chips are still on the wafer one speaks of "Frontend" production; as soon as the wafer is cut into individual chips, the so-called "Backend" production begins. During this stage, the chips are wired and the necessary connections applied. After packaging, the chips are then ready to be inserted into circuit boards.

The manufacturing and assembly process can involve up to five hundred steps. To achieve a cost-effective end product, it is critical that the design of the IC, as well as the manufacturing process, is optimized, in order to reach the highest possible yield of functional ICs from each wafer.

In Dresden in 1999, Infineon became the world's first company to harvest 45% more functional chips from a 300 millimeter wafer than is currently possible from a traditional 200 millimeter wafer – a significant competitive cost advantage for Infineon.

The initial public offering in March 2000 was another important step in the company's history. This made the name "Infineon Technologies AG" very well-known indeed. "Infineon" is a hybrid of the English term "infinity" and the classical Greek word "aön", which means life or eternity. This amalgam signifies endurance, dependability, flexibility, agility and innovation. It reflects the corporate philosophy: "Never stop thinking". Infineon's motto is to always be one step ahead and the strategic goal of the company is to produce the most innovative products in the most cost efficient way. For example, Infineon was the first to produce a biometrical sensory chip in series. The status afforded by intellectual property (IP) is one of Infineon's primary motivations for investing in R & D of IC technology. (The company presently has almost 24,000 registered patents.)

Restructuring for the 21st Century

The public share offer also brought extensive demands for restructuring. Flatter hierarchies and more flexible units were needed to react efficiently to the extreme fluctuations in the highly dynamic market. An organizational structure that could meet the demands of both the customer and market was implemented. Today, the organization is structured into five key business units, or market-oriented divisions. These are Communications & Peripherals, Wireless Products, Automotive & Industrial, Security & Chip Card ICs and Memory Products. Each of these divisions is subdivided into the following components: Front End, Back End and Sales/Marketing.

Going public offered Infineon new opportunities to expand the firm and enter into new strategic partnerships. This has led to increased market potential, since Infineon may now serve competitors of other Siemens divisions as they do any other customer, for example, Infineon may sell its micro chips to Nokia, a competitor of Siemens ICM in the mobile phone market

The importance of Knowledge Management

The above brief description of the current state of affairs at Infineon will help the reader appreciate the increasingly important role Knowledge Management plays at Infineon. Factors both inside and outside the company are reviewed below:

1. *Global competitive forces are increasing performance pressure for Infineon*
   The further development of semiconductor technology is extremely dynamic and competitive pressures are strong. The key factors for maintaining a competitive advantage are performance, productivity and innovations. Infineon has recognized that a crucial factor in this is the organization knowledge, especially that of the employees.

2. *Strategically important knowledge is distributed worldwide*
   The various production sites differed significantly in their levels of performance. Gunther believed that by transferring knowledge gained from experience to other sites, an equally high level of production across all sites could be achieved. Worldwide networks of experts are crucial to the transfer of available, strategically important knowledge.

3. *The production process for chips is highly complex and very knowledge-intensive*
   The production process for chips requires highly specific technological know-how

acquired by years of experience. The speed with which development of new technologies occurs is still increasing, leading to a relatively short "half life of knowledge".

4. *Existing technology is not necessarily a source of lasting competitive advantage*
   It is practically impossible to prevent the competition from quickly imitating, or even improving on new products or methods. Since one core technology is available to virtually anyone, this factor alone cannot ensure a lasting competitive advantage. This has led to the emergence of a global market for ideas where concepts and formulas are freely available to almost anybody.

5. *How much should one share with one's competitors?*
   Strategic alliances and joint ventures have become necessary in order to minimize risks and maximize time and cost advantages. The value of knowledge to Infineon is clearly evident in the form of many patents for new products and production procedures. (These can also be subjected to monetary valuation.) However, questions regarding what knowledge should be shared and what should not, are a central issue in the KM concept.

Knowledge Management at work

Facing a challenge

Infineon's Memory Products division produces mainly memory chips at its production sites across the world. These are found not only in computers, cellular phones, and digital TVs, but also in jets, pacemakers, and ATMs.

Today Helmut Gunther's division is responsible for the improvement of productivity indices of the globally active sites (also called Fabs) of the Frontend. In the early 1990s, Helmut – as responsible manager for transferring processes and for the manufacturing support of both problem-solving procedures and optimizing procedures – had the idea to foster the exchange of process knowledge and to systemize the knowledge. Instead of the current just-in-time and problem-oriented knowledge transfer, he wanted to establish a permanent know-how exchange among the factories, so that they could achieve business goals together.

Helmut was aware of the fact that a crucial prerequisite for the improvement of overall productivity and knowledge sharing would have to be the interconnection of experts. This prompted him to initiate the project KECnetworking (Knowledge Exchange Networking). The project was executed in cooperation with the Siemens AG, Corporate Technology, Knowledge Management Department (ZT IK 1).

In conversations with site representatives, his initial impression that a continuous knowledge exchange promised much potential was repeatedly confirmed. The diverging productivity figures of the different factories, often producing the same chips, confirmed this as well. Furthermore, the process steps for all chip production are very similar and are executed by the same type of machines. This meant that both similar mistakes and development efforts were being duplicated at the various sites. Considering that a single day of missed production, or one missed process step on one machine can cost several million US dollars, the potential for improvement was self-evident.

In the course of conversations with employees, Helmut and his team also gained the impression that the existing databases, which were only constructed for the exchange of documents, weren't a sufficient base for continuous knowledge exchange. Helmut set about motivating the responsible management of the individual sites to take action to improve the situation.

Providing a solution

Initial phase

At the outset Helmut and the site representatives analyzed the situation.

At fourteen different sites over one thousand experts worked in twelve single-process technology areas (see Figure 27). Although these sites did not all produce the same types of chips, there are similarities between the different generations of chips. Experts within the single-process technologies were faced with very similar problems and challenges, it was therefore possible that a site producing a new chip generation could learn from one that had been producing ones of an earlier generation.

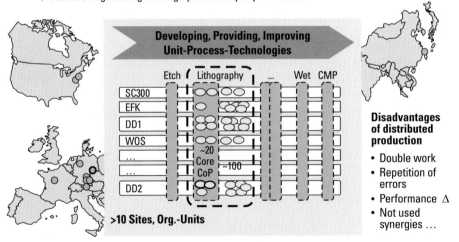

Figure 27 Challenge and Community of Practice solution for KECnetworking

Some interactions had been institutionalized, such as process-transfer projects that have a limited duration, and various regular management committees meetings. The experts themselves could only exchange experiences and provide support on the basis of individual, personal contacts.

Explicit knowledge was documented and exchanged by capturing some of the contents on databases. In most cases the complexity, size, and perishable nature of the explicit knowledge, made this type of documentation process too laborious and not practical enough.

Siemens' rapid growth, its cooperation with other firms, its restructuring, aggressive hiring policy and establishment of new sites had made it increasingly complicated and difficult to develop and maintain such personal networks. In addition, a few cases of repeated mistakes and redundancies happened that could have been avoided by an improvement in the exchange of experience.

Utilizing Communities of Practice

At the end of 1996, Helmut Gunther and the site representatives decided to attempt to improve the situation by interlinking experts in topic-specific Communities of Practice. This notion emanated from Xerox Parc where, at the turn of the decade, the copier engineers had been observed to share their know-how around the water cooler. It has now become an accepted term in the context of Knowledge Management. Essentially, a Community of Practice is a group of people who are peers by virtue of the kind of work in which they are engaged. It is not a formal team, but an informal network, which shares a common agenda and interests. At Infineon, a Community of Practice is understood to be a group of like-minded people, prompted by a mutual interest in a business-relevant area of activity, to share and develop knowledge across organizational boundaries in order to offer mutual support. Through cooperation, not limited by time constraints, the Community has both a virtual and face-to-face character.

The aim was to accelerate the achievement of business goals by promoting the exchange of available knowledge and the development of new knowledge (by improving the networking between experts). The following objectives were formulated for the KECnetworking project:

- Increasing productivity and optimizing production processes.
- Ensuring that process transfers between sites are managed with speed and efficiency to facilitate sharper learning curves.
- Swifter integration of new employees into Community of Practices.
- Reducing the duplication of mistakes at the various sites.
- Reducing redundancies at different locations.
- Better coordination of the process, technology, and facility roadmaps.
- Faster delivery of new, innovative process technologies.
- Improved cooperation with suppliers of facilities and materials.

The first step was the categorizing of the entire Frontend production process into twelve clearly distinguishable topic areas (single-process technologies). For every topic area one moderator from the division Memory Products Frontend Productivity Improvement was chosen, as well as at least one topic representative for each site. Each of these Communities organized an experience-exchange meeting, led by the moderator. These meetings lasted from one and a half to three days, and up to 35 experts from almost every site around the globe participated.

After these meetings, Helmut Gunther and his team gathered feedback on their experience with the Communities-of-Practice approach. The initiative was unanimously approved. The participants had been pleased to have the opportunity to establish good contacts and personal relations with other production sites. There was, nevertheless, equally strong agreement that much could be done to improve on the first KECnetworking activities. Participants complained, for example, that the borders between topic areas were not clear enough, that the focus of the topic areas was too wide, or that the quality of the exchange was not up to standard.

To address these problems systematically, Helmut and the team of representatives decided to analyze their findings. To this end, he assembled a team composed of KEC Community moderators and Siemens internal consultants. Their brief goal was to identify potential areas of improvement, as well as to seek out and implement appropriate intervention measures.

Analyzing phase

In the second phase, the consulting team analyzed the framework and activities of the KEC Communities and the integration of these activities with day-to-day work procedures. Additionally, similar corporate activities and their influences were discussed.

To begin with, interviews were conducted with many of the participants and documentation material and information systems were analyzed. Participation in KEC meetings allowed the consulting team to uncover existing barriers and areas that showed potential for improvement. The results of the first phase of analysis were subjected to a survey of 150 active KEC members, so as to acquire the opinion of and ideas from as wide a range of involved people as possible. The following points were addressed in the survey:

- The individual's perspective on the potential cost-benefit ratio of the KEC activities (compared to the existing ratio).
- The extent to which the individual's need for varying KEC outputs had been fulfilled.
- The individual's opinion on the importance of single-success factors for effective and efficient KEC activities and their degree of fulfillment.
- The type and extent of communication on KEC outputs with site colleagues who had not been directly involved in the meetings.
- The individual's opinion on KEC activities being able to meet expectations and suggestions for improvement.

The analysis phase clearly showed that knowledge management improvement measures were well received by all participants. Core results also showed that there was a definite demand for an efficient and continuous knowledge exchange and collaborative knowledge development. At the same time, the following deficits and barriers were revealed:

- Knowledge exchange by way of semi-annual meetings was not continuous enough. These meetings were largely dominated by discussions of the most recent experiences, while important ones, no longer of immediate concern, were often neglected.
- The meetings lacked goal-orientation and were therefore not very efficient.
- The interest profile of the members was too heterogeneous, as was that of the KEC Community generally.
- The slides presented did not include key lessons. These had been communicated orally only.
- Language difficulties, as well as the use of jargon not known to everybody, got in the way of the exchange of knowledge.
- Between meetings KEC Communities used e-mail as the only form of communication. The existing databases were only used for special projects. A common, specific information platform for each community was missing.
- The distribution of findings from the meetings to other interested parties, who had not participated, was insufficient.
- Experts and other knowledgeable persons were also people who operate under severe time constraints and this meant they could not always adequately participate in the exchange of experiences.
- The participants did not have enough time for support activities, such as the structuring, editing, and maintenance of stored information, in addition to attending community meetings.
- The fear of losing knowledge to joint-venture partners or competitors through closely integrated suppliers, complicated the experience exchange.

The results of the survey concluded the analysis phase and were discussed with KEC moderators and members, as well as the consulting team at a final workshop in March 1998. In just more than a year much had been achieved.

However, if the Community of Practices were to support Infineon in achieving its future goals, further improvement was needed. A concept needed to be developed and implemented that would improve these procedures in the KEC Communities.

Improvement phase

Following the workshops, measures for improvement were defined and concretized, and an action plan was agreed upon.

This included:

- The formation of a KEC support team to aid KEC activities and minimize friction.

- The integration of KECnetworking with the business environment to ensure optimal preconditions for knowledge exchange.
- Improvement of the general organization and modification of individual KEC communities regarding their group composition, the formation of special subgroups, and the coaching by KEC moderators.
- Improvement of the preparation and execution, and a review of regular "face-to-face" meetings and telephone conferences.
- Better structuring of the content of the topic areas and the design of transparent processes regarding experts and sources of information.
- Setting up an information platform, as well as a process for the long-term entry, maintenance, and use thereof.
- Providing appropriate information and communication tools and work methods to facilitate teamwork for groups who are geographically dispersed.
- Internal and external promotional activities to heighten the awareness and acceptance of KECnetworking, as well as to recruit new members.

The consulting team, in close cooperation with KEC moderators and other key individuals, piloted these measures in individual KECnetworking Communities before transferring them to others.

These stages (initial, analyzing, implementing improvements) describe the progress of the implementation of KEC. The following section takes a closer look at each individual component of the KEC, its functions and how it is constituted. These are:

- The Communities of Practice – activities and roles, their composition
- Netmeeting – an online collaboration tool
- An information platform – what it is and how it is structured, the nature of the knowledge shared
- Current activities

Components of the KEC

Community of Practices – Activities and roles

To promote transparency regarding processes and roles in the KEC Communities and to provide specific support for these, a Community model was developed. This model outlined the practical activities of the Community:

- Regular meetings for exchange and for alignment of projects and fab activities.
- Promoting transparency in knowledge areas; providing access to experts.
- Identification, acquisition, and integration of external knowledge.
- Learning from experiences and documentation of explicit knowledge.

- Continuous information exchange and discussions.
- Internal/external benchmarking to locate areas of potential.
- Just-in-time problem solving and mutual support.
- Collaboration and experiments in expert teams.
- Competency development and training.
- Developing new business opportunities.
- The roles identified for these were in essence:

A Community of Practice initiator.

The initiator provides the first decisive impulse for the creation and formation of a Community of Practice. He or she suggests the idea, develops the concept and gives the Community of Practice its initial momentum. He or she can be one of the group's members, but can also be a member of management or one of the persons responsible for Knowledge Management.

A Community of Practice sponsor

The sponsor is generally not a member of the group. This optional role, basically to provide financial and strategic support, can have a strong and positive influence on the success of a Community of Practice.

A Community of Practice moderator/manager

The moderator or manager of the Community of Practice organizes, moderates, and governs the entire group. He or she is the primary contact person and the motivator of knowledge exchange and development (support activities). In large groups, subgroups with their own moderator are often formed. Assistants often help the moderator when he or she provides support activities.

Community of Practice members

The active members of the Community of Practice participate in and contribute to the activities of the Community of Practice. A Community of Practice may have passive members who, for example, are just listed on the distribution list and occasionally use information from the Community of Practice in their work.

Community of Practice support

In companies in which Community of Practices represent a strategic knowledge management instrument, individual Community of Practices often depend on organizational units for support and coordination. This role encompasses easing the burden of operative and supporting activities for members and moderators. The assignments vary from event organization, reediting and researching of contents, to the provision and administration of the IT infrastructure, and the general coaching of the Community of Practice.

Community of Practice external knowledge carriers

The Community of Practice external knowledge carriers with whom contact is made and relationships exist, are very important. They can be internal corporate employees as well as external persons working with partners, suppliers, customers, and – under specific circumstances – even competitors.

The Composition of a Community of Practice

An average KEC Community is more than three years old and comprises 10 to 35 active, and over 100 passive members. To deal with specific subtopics, temporary expert teams are often formed. This approach allows answers to special questions to be found quickly and, enhances the knowledge exchange that occurs between the semiannual meetings of the entire group.

The layering of group sizes – ranging from small expert teams of about five persons to the entire group of more than 100 passive members – has proved to work well for KECnetworking.

While the expert teams develop solutions to specific topics, the active members communicate informally, but regularly, on matters pertaining to the Community and benefit particularly from the experts and the transparency of activities. The passive members generally do not contribute anything to the Community, but wish to be informed about general developments within the Community and are able to communicate informally, if the need arises.

Netmeeting – online collaboration

To support the online collaboration of KEC Communities, experts were provided with the tool MS Netmeeting. Netmeeting is an application that allows users to view and work with the same screen contents on the Infineon-wide Intranet. The use of desktop video cameras, microphones, and speakers significantly improves dispersed teamwork. KECnetworking members are specifically schooled in the use of Netmeeting and provided with a range of aids for the use and monitoring of these sessions.

Information platform and topic structuring

In the beginning KEC Communities did not have access to an adequate information platform which could provide every KEC Community with information that is relevant or important for individual process technologies. In KECnetworking this is done by the means of a LotusNotes database.

The contents of this database are structured in the following way:

• People. Yellow Pages of Community members and external knowledge carriers.

• Teams. Overview of the subgroups, including their topic focuses and members.

• Events. Event calendar with additional information regarding future and past events.

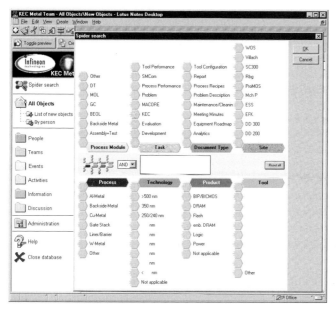

Figure 28 "Knowledge-Spider" as a categorizing and search interface

- Activities. Information about intra- and inter-site projects, evaluations, and process modifications, etc.

- Information base. Additional databases, web links, patents, glossaries, etc.

- Discussion. Opportunity for discussion regarding various topics of the Community of Practice.

Any member can use this information platform to enter or modify information. In Figure 28, a screenshot of the KECnetworking's "knowledge-spider" is shown. The multi-dimensional-content structure facilitates the systematic categorizing of documents and competencies. This mask can also be used to search for information.

Defining knowledge areas

The knowledge focused in the Community is always a critical factor in Infineon's value chain, as it is not openly available on the global knowledge market. Due to the fast-paced and dynamic nature of technological development and the associated decrease in knowledge half-life, a constant need for new knowledge arises. Even if much explicit knowledge can be documented, implicit knowledge is particularly important in the creative and innovative processes taking place.

It is important to recognize that in this process individual participants are dealing with highly specific topics. The number of participants interested in detailed information, or knowledge about any single topic, will be relatively small. For this reason, topics are defined so as to be relevant to all participants, in almost all aspects of their daily job, thus promoting more active participation.

Current activities

Compared to those of the starting phase, current activities are much more intense and multi-layered. The semi-annual meetings that took place in the beginning, have been enhanced by collaboration in diverse expert teams, regular phone calls, information exchange on the Community platforms, and strengthened mutual support.

Now the experts have semiannual to quarterly face-to-face meetings for one central, compulsory module in KECnetworking. These meetings enjoy such high priority among the experts that most of them attend every one, despite their heavy schedules.

During these meetings the flow of detailed information is relatively limited, the focus being on creating transparency in areas of knowledge, identifying mutual interests, as well as building and maintaining personal relations for successful cooperation between themselves. More detailed and concrete exchanges usually take place bilaterally, or in small groups after or between meetings in the long breaks provided. A face-to-face meeting of one and a half to three days has proven adequate.

The participants have explicitly stated that the cooperation between different production sites has significantly improved and they see this as largely the result of the personal contacts forged during the KEC meetings.

The degree of activity in the Community of Practice is subject to strong fluctuations during the annual cycle. This rhythm is determined by the semiannual to quarterly face-to-face meetings as well as by other special KECnetworking activities. At these times, the use of the information platform increases sharply immediately before and after meetings. The relevant employees also invest roughly two full working days in preparation for the meetings. The "continuous activities between regular meetings" that were requested by the participants themselves, have also increased. However, the discussion forum on the information platform remains largely under-utilized for this purpose, with participants preferring spontaneous mutual support or mid-range co-operation in small expert teams.

Outputs

Key outputs of the KEC Communities can be summarized as follows:

- Discovering improvement and innovation potential as well as the transferring and recycling of good practices, insights, and lessons learned.
- Transparency and access to topically relevant and high-quality information from a central source.
- Aligned activities throughout the firm, for example common definitions for technology, facility, and process roadmaps.
- The functional role of the Communities as the technical interface to facility manufacturers.

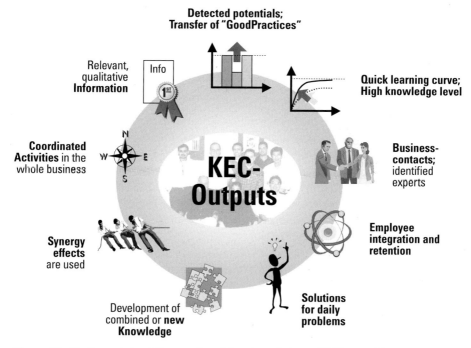

Figure 29 Starting point and Community of Practice solution at KECnetworking

- Development of combined or new knowledge.
- Communal experiments, such as the evaluation of facilities, processes, and new technologies.
- Gaining answers and helpful hints to solve every day problems.
- Integration with the network promotes employee loyalty to the firm.
- Identification of contact persons and improvement of relations network.
- A sharper individual learning curve and generally a higher level of knowledge.

This list of outputs shows how abundant the benefits are that can be derived from KEC Communities. (see Figure 29).

Promotion

Within Infineon, KECnetworking is an innovative and visionary initiative and is therefore of crucial importance. Indicative of this fact is, for example, the fact that KECnetworking is considered one of the firm's 12 globally strategic projects with which Infineon hopes to win the European Foundation for Quality Management's Quality Award 2000. As an example of applied Knowledge Management at Infineon it is also promoted in internal and external publications, not only providing the company with positive public relations, but also strongly motivating KEC members. The name KEC is known far beyond its business boundaries– so much so that an attempt at renaming the project had to be abandoned.

Benefits and success stories

Measuring the performance of KECnetworking activities in an exact manner (for example, calculating Return on Investment) is difficult. So far the creation of a "financially accounted for, internal knowledge market", which is a theoretical option, has not been attempted. Costs can be traced relatively precisely, but benefits are difficult to assess accurately, mainly because they are not easily measurable, and are difficult to quantify monetarily. Many diverse factors make the true influence of KECnetworking difficult to ascertain. To overcome this, an attempt is being made to collect success stories that are directly linked to KECnetworking activities. An example follows in the box below.

> ## A successful good-practice transfer
>
> *Fab A was in need of significantly stronger production capabilities. A new production facility of several million US dollars was considered*
>
> *The moderator of the KEC community took on the role of broker and introduced Fab A's experts to those of Fab B, instead of waiting for them to find one another directly, as is usually the case.*
>
> A detailed comparison with Fab B showed that one process improvement had just achieved a 1.4% productivity improvement at a certain facility. Within two weeks, the process technologists from both sides had transferred the process innovation from Fab A to Fab B. This was no simple task given the different process environments.
>
> The experts confirmed that, in this case, the common context provided by KECnetworking significantly aided the fast and successful process transfer. A substantial investment in a new facility was no longer required and benefits to the company ran into millions of US dollars.

Even if the Return on Investment of the activities cannot be delivered at this point in time, the solid motivation of all participants gives rise to the expectation that in future KECnetworking will contribute positively to business.

The individual participants strongly support these activities and are of the opinion that the benefits resulting from KECnetworking largely outweigh the effort.

Conclusion

Helmut Gunther's past experiences and his vision for the future can be summarized as follows:

The initial success of the project has encouraged him and his team to keep on expanding KECnetworking. The continuous exchange of personal knowledge holds consider-

in research and development. In every corner of the world, in every culture and in every time zone there are Siemens employees. This is a unique source for knowledge sharing. They provide the company with benefits such as the creation and deepening of new knowledge, the ability to detect blind spots within the corporation, and the identification and usage of synergies. Moreover, they facilitate best practice sharing, swifter and more flexible reactions to changes in the corporate environment, the discovery of potential for improvement and innovations, as well as the standardization of terminologies and business processes. For Siemens' employees, Knowledge Communities generate tangible benefits such as swifter access to relevant and qualitative knowledge, and more efficient solutions to everyday work problems.

Considering these advantages, it is not surprising that Knowledge Communities are gaining in importance and are characterizing the way in which innovative and successful companies are organized worldwide.

The lifecycle of Knowledge Communities

In general, a Knowledge Community (KC) is a rather interactive process which forms a certain lifecycle. Such a lifecycle can be divided into three important phases: the start-up, running and improving, and winding-down (Figure 31). Every phase is composed of different activities which can be supported by specific solutions and services. The following description of a Community's lifecycle is especially relevant to Communities which are actively founded and supported, but the ideas are equally relevant to Communities which emerge without any support.

Start-up phase: Establishing a KC first requires awareness through individual motivation, a situation analysis, and a knowledge strategy – these form the initiating process. The subsequent start-up should involve an appropriate member base, as well as a concept of the basic framework in which the KC wants to function. The KCS team has identified the following steps to successfully initiating a Community.

The first step is the preconsideration. A detailed checklist is provided which potential initiators can consult to ascertain the extent to which initiating a Community actually represents an appropriate approach to solving their current business problem. This step is followed by the identification and integration of the type of members that such a Community would require. It is therefore important to survey their common interests and motivate them to actively collaborate on these topics. It is furthermore necessary to

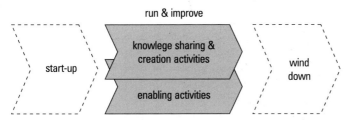

Figure 31 Phases of the lifecycle of a Knowledge Community

find a facilitator who will plan and implement the initiation. The next step is to develop a community concept. This is almost a business plan which can be used to gain the necessary management and member attention. This plan includes factors such as subject, aims, activities, potential benefits, members, financing, organization, and roles. The fourth, and last step of the initiating process, is the setting up of a framework that ensures the workability of the Community. This includes activities to be completed before the start-up, such as informing the management, securing financing, setting up a technological infrastructure, or registering the Community in the Communities@Siemens Directory. Finally the kick-off workshop is the starting point of the Community. At this workshop members get to know one another, develop a joint understanding of the topics to be addressed, plan future activities, agree on common objectives, and organize the structure of the Community.

Run and improve phase: When the KC is up and running, it is critical that continuous improvement takes place. Its members should ascertain that its activities are enabling the creation and sharing of knowledge. Persons with roles to play should be doing so effectively. Having successfully initiated a Community, it is important to keep the momentum in order to develop and sustain a vital and active Community. The Community should become an important component in the everyday work of its members in order to ultimately benefit the Siemens corporation. The community goals, activities and framework must be constantly adapted to the changing business environment.

The KCS team has identified five enabling activities during the running and improvement phase of KCs. Firstly, the organization and moderation activities comprise higher level activities which are necessary for the maintenance and growth of the KC. They can be more content-strategic (i.e. adapting structures), or administrative-organizational (i.e. controlling of group processes) in character. Secondly, the member management and motivation activities comprise activities to identify and recruit new members, and to integrate them into and motivate them for the main activities of the Community. Thirdly, the presentation and promotion activities comprise presentations on corporate events and conferences, articles in employee magazines, integrating the Community into the Communities@Siemens Directory –, in fact all activities aimed at promoting and marketing the Community to relevant stakeholders. Fourthly, the establishing and maintaining of an appropriate IT infrastructure. This is usually outsourced to a service provider. Finally, monitoring and continuous improvement activities comprise activities such as monitoring expenditures or goals development, and constantly adapting and improving the parameters of the KC.

Winding-down phase: KCs should only exist as long as the individual members can see benefits for themselves and their business, and the KC is adapting to the constant changes that take place in a knowledge-intensive environment. If this is not the case, or if the set goals are no longer applicable, it makes sense to discontinue a Knowledge Community. If it becomes necessary to close a Community down, a final KCS workshop will review its activities, processes and outputs. Important aspects of the Community's knowledge are then either transferred to other Communities or archived for later usage or usage, if required, in related knowledge areas.

The role players in a Knowledge Community

A KC can be likened to a drama in which a large number of roles must be assigned to individual players. We will focus on the most important roles that are associated with certain tasks which have to be carried out (Figure 32). The more professionally these roles are performed, the greater the success of the Community. However, one person can cover a number of roles, especially in the start-up phase when the KC is still small. There are three groups of roles in the Community: a group in the Community itself, a group who acts as an interface with the formal organization and a group who performs community service roles.

The Community itself is composed of members – the original actors who generate and share knowledge and the facilitator – the organizer and coordinator of the community activities. The interface roles link the Community to the formal organization and/or other KCs. They are the initiators, the driving forces during the start-up of a KC, as well as being the sponsors, or overseers of budgetary and financial concerns. Finally, there is a group who offers services – usually fee-based – to Knowledge Communities, without being directly involved in the knowledge sharing and creation processes of the Community. The community coach offers methodological and procedural support for all phases of the Community. The IT providers implement and maintain the community platform, tasks which are necessary for an efficient functioning of the Community.

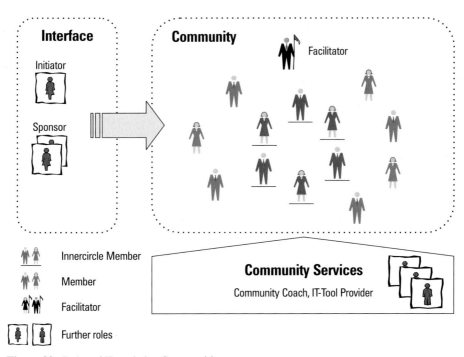

Figure 32 Roles of Knowledge Communities

Communication in Knowledge Communities

Different communication channels can be used for community activities. An appropriate mix of face-to-face and virtual communication is recommended in order to exploit the advantages of both personal contact, and swift, direct communication between the members worldwide. Not only should there be planned and/or regular communication, such as face-to-face conferences and workshops, and phone and video conferences, but there should also be continual and spontaneous communication via phone calls, emails, net-meetings, mailing lists, newsgroups, discussion forums, and information platforms.

Components of successful Knowledge Communities

For a Community to operate successfully, a basic framework is required. The absence of the following success factors usually leads to barriers which, in turn, strongly inhibit the success of the Community:

- people and competencies – the members should have a strong common interest as well as expertise on the subject,
- culture and collaboration – the willingness to exchange knowledge and to collaborate requires trust and a "we" attitude,
- strategy and objectives – the better the core objectives and subject of the KC fit the business goals, the better the potential benefits for the members as well as the corporation,
- knowledge content and structure – the subject of the KC should be focused and clearly structured,
- its processes, roles and organization – a committed, recognized and well organized facilitator as well as engaged, active members are the key success factors of a Community,
- leadership and management support – a Community performs better if it is supported by the management and financially funded by a sponsor, and
- information and communication infrastructure – this is a platform which supports all important communication channels.

Siemens' Knowledge Communities

Before the start of the new millennium, a diverse range of knowledge community projects, such as ICN ShareNet, Best Practice Sharing and KECnetworking, run by individual business units within Siemens, had proven that Knowledge Communities are a very successful concept in the field of Knowledge Management (for more information regarding the above projects, see the following cases in this book: ShareNet – the next generation knowledge management; Practice Exchange in a Best Practice Marketplace; and KECNetworking – Knowledge Management at Infineon Technologies AG).

During December 1999 the Corporate Knowledge Management office (CKM) took the initiative to create a support base for these Knowledge Communities. This support aimed to improve these Communities by providing solutions and services to initiate, run and, if required, wind them down. Ultimately CKM wanted to promote the concept of Knowledge Communities as a new and effective way of collaborating throughout the Siemens organization. To this end CKM approached expert consultants from Corporate Technology (CT), Siemens Business Services (SBS), and Siemens Qualification and Training (SQT) to form the KCS team. This team would initiate and support Communities in order to promote knowledge creation and sharing in and across groups, regions and corporate units within Siemens, as well as to ensure that Siemens is able to access the knowledge of the entire company at the right time. Eventually this mutual effort would evolve into the company-wide Knowledge Community Support.

The development and implementation of KCS

The kick-off

Before December 1999 there had been an effort to support Communities through a number of different community projects and activities throughout Siemens, but in most cases these efforts were restricted to single corporate units. This experience had, however, been meaningful and successful. In December the question was how to actively and systematically support Knowledge Communities with important subjects and themes, and how these Communities could be improved in order to function more efficiently and generate more benefits for the business.

What kind of approach was needed to fulfil the above objectives? There was simply not enough consulting capacity and too many themes or subjects to individually and personally coach and support every single Community within Siemens. In December 1999, in an initiative driven by CKM, consultants from the different departments came together to think about a central, company-wide, professional and comprehensive Knowledge Community Support. It was agreed that a joint approach was necessary. The departments would pool their competencies and be partners in this endeavor.

They decided on a support approach that had been successfully used in project management, a similar support field in which it is also essential that the manager is skilled and trained. To this end trainings and support materials are developed to facilitate the management of a project. This was an approach that KCS wanted to follow as well. The idea was that, on the one hand, the KCS team would personally provide fee-based support to highly business-relevant Communities and, on the other hand, the numerous other Communities which had emerged would be supported by the KCS by providing them with trainings, support materials, know-how and a technical infrastructure. The goal was to provide as much know-how as possible so that the Communities could do as much as possible themselves.

At that time the team defined a mission for the KCS: to support Knowledge Communities throughout their lifecycle; to facilitate the identification of Knowledge Communities; to facilitate the transformation of Siemens into a global network of knowledge and

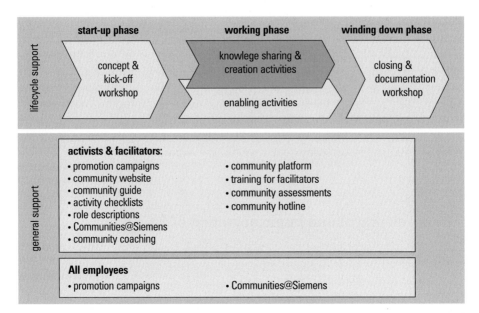

Figure 33 Solutions and services of the Knowledge Community Support

innovation, and to promote cooperation based on trust, mutual respect and open communication.

During the period January to end April 2000 the KCS team developed a concept and a a roadmap for KCS, in order to argue for the envisaged support of the Knowledge Communities. An important task was to create top management awareness of Knowledge Communities within the Siemens Group. These Communities had to be recognized as an important, strategic issue with a high business impact that should be integrated into the business processes and evaluation mechanisms. It was also crucial that the top management truly supported these activities beyond mere lip service.

Parallel to their activities to obtain the management's attention, the KCS team started developing the first solutions and services. These support solutions and services were conceptualized to be either free of charge (e.g. material on a website), or be charged per service provided (e.g. community coaching). There were two perspectives to these solutions and services. The first perspective was that some of these solutions and services could be directly linked to a certain phase in the lifecycle of a Community, and therefore would only be offered or presented during such a specific phase. The second perspective was that the remaining solutions and services would be of a more general character which would be required by Knowledge Communities throughout their lifetime. These two perspectives are illustrated in Figure 33.

The activities of the KCS team can be grouped into six action fields which are described in the following:

- to launch the KCS website in order to provide "help to self-help" for the whole KC lifecycle,

- to promote KCs and their support within Siemens,
- to install a KCS Hotline for the activists and facilitators,
- to provide KC coachings jointly (previously each department represented on the KCS team had done this individually using its own methodology),
- to design and implement a Communities@Siemens directory, and
- to create detailed activity and role descriptions for Communities

KCS website

For the website the KCS (Figure 34) team gathered the know-how of and experience with the support of KCs that were present within Siemens. This knowledge was documented on the KCS website and provides community activists and facilitators with detailed information on what is important throughout the life cycle of a Knowledge Community, what should be done and what should be improved within the Community. If an individual has questions or comments regarding KCs, ranging from the most basic "how-to" to the more complicated "why" questions, a good starting point is always the KCS website. This information comprises a wide spectrum of useful knowledge, for example practical cases of Communities within the Siemens Group, definitions, tips and tricks with which to start and further improve Communities, and a library with a plethora of information (e.g. internal and external presentations, books, articles, web links). Also included in and described on this site are fee-based solutions and service offerings by the KCS team for the successful initiation and running of Communities. There are also links to the providers of standardized intranet collaboration platforms which can be used for knowledge community purposes and which have proven to be of much practical value.

Figure 34 The Web-site of the Knowledge Community Support

In October 2000 a second, completely revised version of the KCS website was launched. All the know-how gained in KC coachings and other support activities had been included. This time the website was provided in English and German.

KCS promotion

From a promotional point of view the KCS activities were aimed primarily at the community activists and facilitators, and more generally at all Siemens employees. The team's goal was to help Siemens develop their KCs with maximum efficiency by helping them to get to know and utilize the service offerings and support material of the KCS. It was surmised that such an approach would motivate other employees to become activists and initiate Communities of their own.

The goal of the KCS team when focussing on the general company employees, was to have all employees know what a Knowledge Community entailed and what benefits Communities offered their members and the corporation. Moreover, all employees had to know where to find existing KCs, or where to find information on them. To reach these goals the KCS team, at the end of May 2000, launched a first promotion campaign to create awareness and to identify existing Communities. Simultaneously extensive interviews were held.

In December 2000 the KCS team started the second extensive promotion campaign. They promoted their activities via internal journals, newsletters and staged presentations at internal events. CKM also conducted a large internal survey on Knowledge Management during which more than 7500 employees were surveyed. A part of this survey dealt with the issue of Knowledge Communities. The results of all these activities were used to improve the KCS.

In July 2001 the development and publication of the Knowledge Community brochure for community activists, facilitators and sponsors have been completed. As a derivative of the extensive knowledge in the KCS website, this brochure gives a detailed, but concise overview of all the important aspects of Knowledge Communities. Other professional promotion material is also under development.

KCS hotline & community coaching

The KCS Hotline is another service developed by KCS. Activists and facilitators can send emails outlining their questions and/or needs to this hotline, which will be answered by the KCS Hotline team. This team is composed of consultants and experts from CKM, CT, SBS and SQT who have many years of experience in knowledge community projects, both within and outside Siemens. Members of this consultant network maintain close contact with one another, and will pool their particular skills and experiences in order to provide facilitators, initiators and sponsors with the best possible solutions.

During the start-up phase of a Community it is, however, not always easy for the initiators, sponsors and facilitators to know exactly how to proceed. To overcome this problem, the KCS team offers a community coaching service. If required, such a coach provides the Community with methodological advice on a fee basis, but does not involve himself with the knowledge sharing and creation processes of the Community.

Such services include staging of presentations for management, drafting community concepts, providing IT tools, designing, moderating, and organizing events, and conducting a community assessment. A Knowledge Community is only as effective in creating and sharing knowledge as its facilitators and members are. The KCS team has therefore developed and piloted a training program specifically dedicated to train KC facilitators.

Communities@Siemens Directory

October 2000 saw the launch of the Communities@Siemens Directory (Figure 35) providing details of existing Communities. The promotion campaign and active search had succeed in populating this directory with existing Communities before its actual launch.

The Communities@Siemens Directory is a tool which forms part of the Siemens Employee Portal, but which can also be entered via the KCS website. It offers Siemens employees a quick and user-friendly way of finding a Community on a certain topic, and information about the Community. Besides an index and a full-text search, this directory is subdivided into a "topic tree". Currently Communities on topics such as business processes and functions, and working methods and tools can be found.

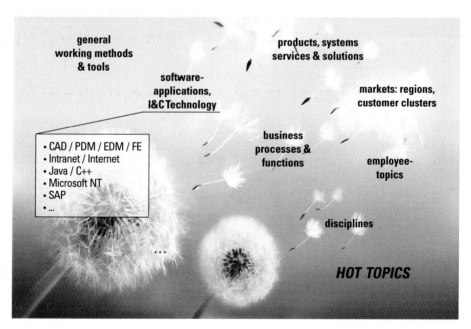

Figure 35 The Communities@Siemens Directory

Every new Community is strongly encouraged to register in this Communities@Siemens Directory in order to allow potential new members to find it. Registering requires that the facilitator, or sponsor of the Community answers a few standard questions regarding the name and subject of the Community, the description of the target group, the working language, the types of activities, the names of contact persons, and the website address.

KC support material

In September 2000 the KCS team started creating detailed checklists, roles and activity descriptions for almost everything related to Knowledge Communities. These lists and descriptions include issues such as: general cost and benefit overviews, templates and descriptions of the different knowledge sharing and creation activities, checklists for enabling activities and community roles, and some hints on how to wind a Community down. This support material is further supported by a glossary and a list of frequently asked questions on the KCS website which explains all terminology and answers all important questions that Community newcomers may have. This activity is especially important in fulfilling the original goal of providing Community activists and facilitators with as much know-how on KCs as possible.

During the start-up phase of a new Community, a detailed to-do list, or some type of recipe for success is extremely helpful. A number of checklists are provided by KCS to guide a community initiator or facilitator as painlessly as possible through the necessary steps. The checklist includes certain approaches to solving problems that may arise, and simultaneously serves as an education medium about the general nuances associated with Communities at Siemens.

Key Propositions

1. For the Knowledge Community Support, the individual approaches, concepts, views, ideas and methodologies of different departments have been brought together. This was, on the one hand, a challenge requiring much effort to reach agreements and develop a common understanding, goals and methods of working. On the other hand, this was a rewarding, creative and interdisciplinary effort involving more manpower than any single department could invest. KCS is also more stable and flexible in the face of the turbulence facing business today.

2. Both the implementation of a community IT platform (Siemens ShareNet) and the development of the employee portal have been strong enablers in aligning the numerous existing Knowledge Communities within Siemens and the activities around them.

3. Success stories such as the ICN ShareNet have proven of great importance in obtaining management support to start a Community Support. It is far easier to obtain the necessary resources if community concepts are already working and producing benefits for the company.

4. Within Siemens there are countless Knowledge Communities that are often not called such, or that have been set up for other purposes. This makes it difficult to identify Knowledge Communities and it is therefore very important to approach activists within Siemens to register their Communities in the Communities@Siemens Directory.

5. The issue of supporting Knowledge Communities is rather facile on an abstract or theoretical level, but supporting them on a concrete level is no longer elementary. This is highly relevant for the descriptions, checklists and other support material the KCS created to support the Communities. They are now in a version which is well understood by the activists and facilitators, but to create support material which can be easily understood by all employees is still a challenge to be faced.

6. To individually coach actively founded Communities and to provide all others with support material have proved a successful approach. The activists and facilitators of actively founded Communities demand and deserve coaching and training, however Communities which emerge spontaneously usually want to work on their own, but are nevertheless very happy to receive checklists and activity descriptions.

7. Creating support for Communities on an incremental way was another successful approach that allowed KCS to quickly set up a website and offer some services. In time everything has become more sophisticated. KCS envisage that in future community management will be as common as project management.

Discussion questions

1. How could one promote the idea of Knowledge Communities as a new way of collaboration within such a large group such as Siemens? Keep in mind that professional support is available for these KCs.

2. Which elements of the Knowledge Community Support have probably had the most impact on Siemens?

3. How can one populate and improve the Communities@Siemens Directory?

4. What are the important steps and issues in the start-up phase of a Knowledge Community?

5. What are points for improvement of the KCS concept?

6. What are the advantages and disadvantages of supporting Knowledge Communities with the help of several departments?

IV   Added Value of Knowledge Management

A guided tour through knowledgemotion™: The Siemens Business Services Knowledge Management Framework

Tanja Gartner, Hans Obermeier & Dirk Ramhorst

Abstract

This case study illustrates the Knowledge Management Framework that was designed for the introduction of knowledgemotion™*, the company-wide knowledge management initiative at Siemens Business Services.

The Knowledge Management Framework will give the reader an understanding of the integrated approach to Knowledge Management and the different stages of implementation. It also introduces the key learning processes experienced by Siemens Business Services during the various implementation phases of knowledgemotion. The knowledge management (KM) requirements, challenges and solutions within the service business are highlighted.

The case study also shows the requirements, objectives, challenges and solutions of (KM) programs, in general, and at Siemens Business Services, in particular.

Based on the KM implementation experience at Siemens Business Services, the case study closes with critical success factors for KM implementations, both within and outside Siemens.

Introduction

In the fiscal year 2000/2001, Siemens Business Services with its more than 33,500 employees in over eighty countries around the globe plans a turnover in excess of 5.8 billion Euro. It is a leading full-service provider, supplying consulting services, system integration, operational services and outsourcing on an international level.

Siemens Business Services was the first Siemens unit dedicated exclusively to a service business. Since it only sells the knowledge of its employees directly and no other products or systems, there has always been a clear understanding of the importance of a

---

* knowledgemotion™ is a registered trademark of Siemens Business Services in more than 40 countries.

people-oriented knowledge management (KM). However, various groups within Siemens Business Services work in the field of IT tools, which means that IT – not people – continued to be the focus of many discussions about the KM program.

Besides the focus on IT, the Siemens Business Services KM team faced the challenge of uniting over 40 separate "knowledge islands" within Siemens Business Services which had been dealing with KM in some form or another, using different technologies, structures and processes. The Siemens Business Services approach to KM had to be a common one, keeping in mind that only an integrated concept would allow the employees to exchange their knowledge across organizational boundaries in order to leverage local knowledge globally. This vision was set and a common name for the Siemens Business Services-wide, global Knowledge Management Initiative was found: knowledgemotion. Move the organizational knowledge with paying attention to the emotional side.

In parallel with the internal activities, the Siemens Business Services KM team combined KM competencies from the technical side with those of the management-consulting, change management, communication and HR side in an Siemens Business Services-wide knowledge management Community.

In the many discussions that followed, both internal and external, the team learned exactly how difficult KM issues are to discuss. The whole field of KM is very complex and has many links to other areas, such as human resources, project management, corporate marketing, workers' councils etc. Slides – many of them – were needed to explain the team's understanding of KM and its approach to it. It was rather like building the tower of Babylon.

In order to facilitate those KM discussions, the team developed a KM framework that allowed the creation of a common understanding through the use of only one diagram, rather than a whole set of slides. Using this framework, knowledge management at Siemens Business Services aimed primarily at developing an understanding and a transparency of the knowledge and competencies of the employees. The next step involved individual and organizational learning through experiences with projects as well as a corporate knowledge base which allowed the capture and re-use of knowledge assets in the Siemens Business Services business processes.

When Siemens Business Services was established in 1995, a number of its core units emerged from units within Siemens Nixdorf Informationssysteme (SNI). Within those units, SNI's CEO, Gerhard Schulmeyer, had successfully pursued an extensive process of "culture change". Consequently it was possible to base KM at Siemens Business Services on a corporate culture that was already familiar with and receptive to the subject. In contrast with their competitors, the challenge of knowledgemotion did not involve finding a "classic" consulting culture as the basis for KM. The challenge involved changing a corporate culture that was still strongly shaped and molded by past products and industries.

The aims of knowledgemotion and the general cultural conditions supported the view of an integral KM approach combining a KM organization with defined KM processes and roles as well as change management activities. This fostered a lasting impression that the specified objectives could not only be achieved by a KM initiative that was mainly tool-oriented.

Tool-oriented initiatives often take the form of "stand-alone" solutions which rarely meet the requirements of the business, and often result in serious resistance by both management and employees. Furthermore, many tool-oriented KM initiatives reveal few benefits to motivate their users.

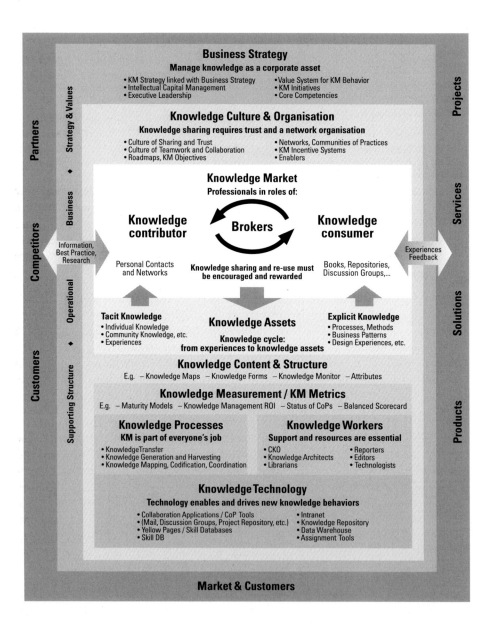

Another challenge facing Siemens Business Services was a rapid and strong organizational growth – as high as thirty percent in some units. The fundamentals of KM also needed to be imparted to the new employees to lay the foundation of a knowledge-based culture. The consistency of a KM organization would, moreover, allow a smooth transformation in times of restructuring.

A description of the KM framework is provided as a guide to presenting the individual methodological modules that form the complex world of knowledgemotion.

Introduction to the KM framework

The objective of the KM framework, as mentioned earlier, is to develop a standardized KM understanding based on the presentation of the dimensions and operations of an integral KM model. Its special feature is that it is compressed into one single image. The framework is used both internally and externally as a method of describing KM instruments and judging the comprehensiveness of customers' KM programs and the way in which the instruments interact.

The framework must be read from the outside in for a complete understanding of this KM approach.

The core of the framework describes the basic principle of successful KM: the knowledge portal where knowledge contributors and knowledge consumers meet.

With this background in mind, we can start on our guided tour of the ten steps.

1. Knowledge management meets business strategy

An important success factor in designing KM programs is their relationship to business strategy and their integration into the core business processes.

In concrete terms the business strategy for KM at Siemens Business Services involves better equipping other service units to fulfil the needs arising from their interaction with Siemens and external customers. In the event of above-average growth, the KM strategy will allow Siemens Business Services and the corporate culture to keep pace with this growth.

On the one hand, Siemens Business Services, having grown out of SNI, has already successfully completed the paradigm shift from a product-oriented to a service-oriented company. On the other hand, KM for Siemens Business Services as a pure service company, relates directly to intellectual capital management in that its corporate value is not reflected by the balance sheet value of, say, the fixed assets, but is based primarily on the intellectual assets. These assets, in turn, are represented by the processes, the structures and the relationships Siemens Business Services has with customers, partners and employees.

From this point of consideration, we derived the guiding principle:

Manage knowledge as a corporate asset and
leverage local knowledge globally.

From the strategy used at Siemens Business Services and in the individual units as carriers of KM, the team then developed the objectives and the timeframe for each of the steps in the form of a roadmap. In so doing, it was important to consider which modules were already in use and only needed to be developed further – specifically for knowledgemotion.

This strategy had two important results:

- the enhancement of individual KM initiatives that were designed to add value to the relevant business areas, and
- the description of the core competencies.

This meant that Siemens Business Services had to concentrate on KM as applied in projects.

On our tour this strategy forms the framework for further KM activities and we can now move on to the second step.

2. Knowledge culture and organization

Of all the elements integrated into the KM strategy, it is the corporate culture and prevailing company values – in particular – that determine the success of a KM initiative.

In order to evaluate the initial situation within the organization, the team carried out interviews and surveys to get an idea of how ready employees were to accept knowledgemotion. In this context the team discovered that the values of "sharing" and "trust" were particularly crucial for the success of KM. More than 50 percent of the interviewed employees mentioned that the organizational structure must be designed to foster the generation and exchange of collective knowledge. A "knowledge is power" mentality would only lead to competition between departments. This posed a huge challenge, one that still exists, to maintain KM values during times of rapid organizational growth and constant internal development. Given the importance of project work at Siemens Business Services, the culture of teamwork and cooperation within the organization are also extremely important for the success of KM activities.

If you analyze the status of an organization with a view to implementing KM, an important question is: How can the employees be motivated to share knowledge? In the long term, the exchange of knowledge should ideally be a basic concern of every single employee (an intrinsic motivation), as is already the case with some consulting firms. The regular communication of goals and benefits, frequent KM activities, promotion of success stories, teambuilding measures and last, but not least, commitment and active support by the management, help to encourage this motivation. However, in the short term the team had to resort to using extrinsic motivation, introducing KM into the general job description, the staff dialogue etc. An example of this would be when employees get a certain percentage of their variable income based on KM results (sharing and re-use of knowledge). However, knowledgemotion, enables employees to earn additionally "shares" through their contributions. These "shares" accumulate and can be converted into knowledge-related rewards, such as technical literature, training courses or communication tools, e.g. mobile phones.

The results of the initial research indicated that time was seen as a demotivating factor, i.e. a crucial obstacle to KM. The team therefore tried to design "windows of time" in which employees could exchange knowledge, or create new knowledge within a specialized subject area. The extent to which this goal has been achieved is measured in terms of a cost-benefit relationship. By successfully satisfying these requirements, employees can determine a part of their variable income.

KM has also been incorporated into the annual staff dialogue, during which it is possible to assess how an employee is getting on with knowledge as a resource (among other things). This annual event is part of a 360° feedback from colleagues, superiors, customers and partners which can also result in the implementation of certain measures, including changes in the employee's future career plans. In this way incentives are not only used in a positive sense, but also negatively when they lead to the development and enforcement of sanctions, e.g. preventing employees with low ratings in the KM evaluation from reaching senior management positions.

A particularly important property of every KM reward and recognition system is specific support for the relevant KM program by making adjustments to the program as problems occur. In the long term, however, the design of the KM program should persuade employees of its benefits. In other words, they should be encouraged to exchange knowledge not only for monetary rewards, but simply because they are convinced that it is a good thing in itself.

In addition to the incentive systems, the team also had to consider the organizational implementation of KM. This involved, first and foremost, the networking of employees within the organization, as well as support for employees in projects by, for example, setting up a central knowledge center.

At Siemens Business Services the networking of employees to facilitate the exchange of their knowledge, is self-organized in "Communities". Communities are groups of employees with a common interest in a business subject matter and/or similar tasks. The members of a Community exchange business-relevant information and experiences, discuss current topics and work on knowledge projects across organizational boundaries. They are open to every employee and can be launched by anyone. A precondition for the launch of a Community is a business-relevant purpose. This could be, for example, the creation of a new service offering, the optimization of portfolio elements, mutual project work, or reciprocal training in a current topic.

In addition to the virtual cooperation between the members of a Community on the collaboration platform of the knowledgemotion system, Communities at Siemens Business Services regularly meet face-to-face. During these meetings the members focus on the exchange of knowledge, for example, on project experiences, or on working out new facets of the topic. The central objectives of the Communities also include the incorporation and subject-specific induction of new employees.

At Siemens Business Services the Communities that attain the highest maturity level are those in which the corporate knowledge and competencies are bundled into basic topics. This is worth noting, because this model was introduced using Siemens Business Services' organizational matrix. This means that employees from all relevant corporate units and countries work together in Communities to offer the market these basic topics.

The work in the Communities is very highly institutionalized. There is a core team, whose members devote more than half of their daily work hours to Community work. There is also an operational team whose members invest up to a quarter of their time to community work. Then there is a full team who do not have to satisfy any specific requirements regarding time. Membership of a Community is voluntary. Forums are available through knowledgemotion for successful – and virtual – cooperation. The forums will be discussed in more detail under knowledge technology (step 10).

Against the backdrop of these strategic, cultural and organizational conditions, the core of the Siemens Business Services KM model, namely the market principle, will be examined in the next step of our tour.

3. The market principle

In an organization like Siemens Business Services, there are both "knowledge contributors" and "knowledge consumers". The contributor and consumer model does not mean that their positions are fixed. Depending on the topic and the work situation, every employee acts both as a contributor and a consumer. Within the framework, they are therefore referred to as *professionals in the role of* a knowledge contributor, for example.

The initial situation at Siemens Business Services could be described as being an unstructured knowledge market. Employees did not know who had performed a business process optimization in a certain area within a car manufacturing company, or who had other experiences to offer. The reverse was also true. There was very little transparency about those employees who had perhaps canvassed the same customer for business in the same industry and what the status of this activity with this customer was, and what kind of business this entailed. And, of course, there was no chance of developing or transferring a specific methodology between the various teams.

The knowledgemotion portal at Siemens Business Services now offers a means for both knowledge contributors and consumers to exchange their knowledge on a clearly structured and open market place.

4. Knowledge managers and knowledge brokers

One of the first solution modules dealt with establishing knowledge managers and knowledge brokers. Knowledge managers ensure that the processes and the technology of knowledgemotion are constantly used to ensure that the knowledge in their units or countries is accessible Siemens Business Services-wide. In addition, they supervise the efficiency of the KM processes as well as the achievements of the relevant improvements made in their fields.

Knowledge managers, together with the unit/country managers, align the KM activities with the Siemens Business Services targets and are the drivers in their respective units of Siemens Business Services' conversion into a knowledge-based company. They promote knowledgemotion by providing a KM-friendly environment. Besides their expertise knowledge, knowledge managers also apply change management methods.

An expert in their respective topics, knowledge brokers are familiar with the core Siemens Business Services processes and the projects in their environment. They are in close contact with their units' employees and the Communities that work on topics in their field of interest.

Knowledge brokers

- are the first contact for all questions regarding Knowledge Management;
- train, motivate and support the users of knowledgemotion;
- administer the Siemens Business Services knowledge base – by checking the validity of new contributions according to set criteria and turning these submissions into "knowledge asset candidates", and remove out-dated content;
- generate additional "knowledge asset candidates" from the content of the project bases and the know-how of experienced employees;
- migrate suitable contents from KM "island" solutions to the Siemens Business Services knowledge base;
- ensure that a Community has an opportunity to do a quick content check of a "knowledge asset candidate" to ensure that the quality of all knowledge assets have been thoroughly checked;
- answer their unit users' FAQs (frequently asked questions) and publish them in the knowledge base;
- identify potential Communities, support existing Communities and help employees with the founding of new ones;
- administer the incentive system of their unit's knowledge contributors;
- collect feedback and suggestions; and
- collect and communicate news and success stories.

During the implementation phase the team was faced with the decision of whether to introduce *full-time brokers* into the organization, to use the knowledge brokers as *a function*, or define being a broker as *a role* (i.e. an employee spends part of his/her workweek as a broker). After weighing up the options, they opted for the role model, since the interaction between daily business – the direct source of knowledge – and its management promises to add significant value.

Special training, strategic participation in projects and networking in a separate Community provide the designated knowledge brokers and knowledge managers with the experience they need to fulfill their roles.

In addition to the knowledge management roles, all the other people in the organization should also assume the role of "knowledge brokers". Anyone who is confronted with a query to which he or she does not have a simple answer, but knows someone who can help, acts as a broker – in the sense of competency – by establishing the connection between the problem and the solution. To put this approach into practice effectively, participants should be motivated to share and use knowledge supplied by others. Once again the underlying importance of a knowledge culture is obvious.

5. Goods available at the knowledge marketplace

In the early stages of the implementation of knowledgemotion, it soon became evident from the interviews with Siemens Business Services employees that both implicit knowledge (competencies and experiences stored in the heads of employees) and explicit knowledge (knowledge stored digitally on hard disks, file servers etc.) would be exchanged. The question was: How can implicit knowledge be exchanged unless it becomes explicit? A way of doing this is to personalize documents and thereby link readers to the author who serves as a source of additional information, competence and personal experiences. Another possibility is a system of "urgent requests" that are posted to a digital notice board and bring employees facing a specific problem in contact with competent colleagues. In addition to the knowledge held by an individual, other important components include the knowledge contained in a project team, or in another group working in close cooperation, mostly represented by Communities.

As mentioned before, explicit knowledge comprises documents, processes, methods, business patterns, and so on, and must therefore be dealt with in a much more concrete way. This in no way simplifies matters, since we are, for example, still faced with the question as to the maturity level of a knowledge "asset" (i.e. document). This is basically determined by the degree to which the knowledge can be re-used – the higher the degree of reusability and reliance, the higher the value of a knowledge asset. This poses a further challenge: how to extract the lessons learned from a document, since just a few projects could lead to thousands of documents. A search on the document if it were presented as text, would not lead to a useful result – it would be like an unspecified search on the Internet. One would not be able to judge the document quality in detail, or assess the quality and maturity of a specific checklist to determine if the policies used proved to be good or bad practice. And more importantly, there would be no way of refining this knowledge.

In order to trade the documents on the knowledge market, we have to learn from experience, and identify and use knowledge assets in a methodical manner. Therefore explicit knowledge has to be quality-checked, improved and altered in defined process steps (see step 8) into knowledge assets.

Siemens Business Services defines knowledge assets as the elements that represent its knowledge and experience in value-added processes. These elements can be examples, checklists, case studies, templates, architectures, business frameworks, practice guides or even a methodology component developed from practical experience gained in projects.

However, even with validated knowledge assets, which are stored separately from non-validated project documents, a full text search, for example, is still only partly success-ful. In knowledgemotion, the content structure of the knowledge base reflects the areas of knowledge along the core business processes of Siemens Business Services that pro-vide a quick and structured access.

6. Knowledge maps

Another content-related element was, for example, the development of the methodol-ogy for creating knowledge maps – not in the sense of Yellow Pages, as they are often understood, but the graphic display of knowledge flows and competency networks.

Different colors describe various competency implementations, while connectors show the intensity of the knowledge flows. The size of these networks is shown, the inter-faces to partners, and special, possibly critical node points in the organization. Knowl-edge maps have made the implementation of expert networks (Communities) in organi-zations easier.

7. Knowledge measurement and KM metrics

The purpose of the models developed in this complex topic, is to determine the KM maturity level of an organization, the Return on Investment (ROI) of KM programs, and to measure the success of KM projects and approaches to intellectual capital man-agement, using Balanced Scorecards.

8. Knowledge processes

The integration of KM into the corporate processes is a critical success factor for every successful KM implementation. The differentiation of the flow of tacit and explicit knowledge lead the project team to look at and extend numerous processes within the context of KM at Siemens Business Services. Let us examine a few of these basic knowledge processes.

In order to gather and capture knowledge in the daily business, the project team defined *a knowledge lifecycle process*. Knowledge contributors publish documents which might be useful to their colleagues in the "knowledge upload area" of the Knowledge base. After a formal check by the knowledge broker, the documents obtain the status of a "knowledge asset candidate". They are moved to the "knowledge assets" area and can be re-used by Siemens Business Services employees worldwide. In addition, knowl-edge asset candidates which have been content-checked by the experts of a Community dealing with the respective business topic, become "knowledge assets" and can be con-sidered best practice.

This process, however, needs "owners" to encourage the process, and roles must be allocated, each of which has a specific responsibility for output. The knowledge broker, as discussed earlier, acts as the process owner. Once documented knowledge has been verified by this process, the quality and maturity thereof are considered proven. Knowl-edge assets and knowledge asset candidates can then be reused in other projects, thereby integrating them into a cycle of constant reuse and further development. Knowledge assets are also important for Siemens Business Services' own business development. Through expert counseling, or by providing clients with a case study, knowledge assets are used to present expertise in a specific field.

In order to make implicit knowledge (for example, lessons learned) that has been acquired in particular projects, available to other colleagues, and to identify the knowledge assets, the team defined a *project debriefing process*, conducted in the form of workshops, for the systematic evaluation of individual and team lessons.

This was integrated into Siemens Business Services' project-delivery process, and is now part of the project process. A project debriefing workshop is conducted either when a specific project milestone is reached, or at the end of a project. It examines the relations with the customer and the development of the individual project employees, as well as the collective learning results. This structured debriefing makes a huge demand on the corporate culture, so that everyone involved may learn from problems encountered in the course of the project. The questions asked at a debriefing relate particularly to the progress of the project and possible deviations from the project plan.

The project leader is responsible for initiating this project-debriefing session. The participants are: the project leader, project members and a Community representative (or another member of the Community). The workshop is facilitated according to a predefined structure which can be tailored to the specific circumstances of the project. The objectives of this workshop are to:

- review the approach chosen for the project (what was conducive and what was obstructive to the project delivery process)
- review the project results in terms of the business value achieved
- identify lessons learned and best practices in the project
- draw conclusions and develop measures to help repeat past successes and make improvements for future projects
- document the experience, and convert implicit knowledge to explicit knowledge
- identify knowledge asset candidates.

The results of the project debriefing are documented and contributed to the knowledge base in order to make them available for Siemens Business Services colleagues worldwide. The workshop itself is not open to everyone, as it was found that conflicts or bad experiences within the project team (if there have been any) are better resolved in a confidential setting.

9. Knowledge workers

To promote the KM process, the project team defined other roles for KM workers beside those of knowledge broker and manager. These roles were mostly defined within Communities.

Community brokers are the driver of the development and exchange of knowledge in a Community. They direct the Community, coordinate and organize the activities and resources. Community brokers have subject matter-specific methodological and technical knowledge, but do not need to be an expert., They should preferably use their social and communication skills to motivate members to participate actively in the Community. In order to distribute the work load, the community broker role can be assigned to several people.

Community members form the basis of the Community. Active members regularly participate in the activities of the Community, create and exchange tacit and explicit knowledge. Passive members are merely included in the Community's distribution list and occasionally use information from the Community. The community members may act as consumers and as contributors of knowledge. With regard to their personal knowledge, community members assume the role of subject matter experts (SMEs). They contribute knowledge and experience on a particular subject, e.g. industry, solution, product, or technology. Normally SMEs are successfully involved in a number of engagements and are highly regarded by their colleagues.

10. The role of technology

The description of the other parts of the framework may underscore the importance of the right cultural environment for KM, but the special role that technology plays should not be overlooked.

When studying the time frame needed for the transformation of a corporate culture into a knowledge-oriented one – where knowledge sharing and the collective creation of new knowledge is part of everyday work – it is clear that this takes a long time. On the other hand, the successful provision and introduction of an intranet-based tool to facilitate the process can be achieved in a few months or less.

Similarly, if you look at the possibilities for the exchange of knowledge within global organizations through Internet/intranet technologies, two important roles can be deduced for technology: being drivers and being enablers of KM.

A successful KM program must therefore plan the individual program components in such a way that each employee recognizes the personal benefit he/she could obtain from KM in the overall architecture (e.g. through the provision or optimization of a tool). Technological solutions support the success of the entire program and can thus be used as "quick hits" in KM projects. (Only a brief overview of this topic is given, since the description of the technologies used in KM falls outside the scope of this case study.)

The technologies of the knowledgemotion system include the following:

1. *Knowledge portal*: The Knowledge Portal is a central entry point to the corporate knowledge. It combines the knowledge base, ShareNet (see below) and different project bases (see below) under one user interface. This means that the user only has to log in once in order to have access to all applications (single login). The portal appearance can be adapted to the preferences of the user and includes an efficient search engine providing well directed access to the Siemens Business Services knowledge.

2. *Knowledge base*: The knowledge base contains the knowledge assets, i.e. the quality-checked explicit knowledge of the organizational units of Siemens Business Services and of the Communities. For example, templates, presentations, methods, samples and experiences from successful projects, or market, industry or client information. Each document has a defined author or owner who is simultaneously the contact person for further questions regarding the content. A short English description of the content and attributes guarantee a global use of the knowledge assets. The access takes place either via a form-based search or via browsing through a standardized content structure corresponding to the Siemens Business Services business processes.

3. *Project bases*: Besides the knowledge base, several operative units use a project base which documents current and completed projects in the form of proposals, concepts, methods or debriefings. The project bases are accessible via the knowledge portal so that all employees can profit from their content. Selective access rights ensure the security of the data.

4. *ShareNet*: ShareNet is the platform for the global communication and collaboration of all Siemens Business Services employees. Communities meet in the discussion forums, cooperate on specific business tasks and document tacit knowledge. The latter provides the basis for knowledge assets which are moved to the Siemens Business Services knowledge base.

5. *Method base*: The method base contains the methodology of Siemens Business Services documented in the so called Chestra (see Knowledge Management for the e-business transformation) handbook. Through the method base, employees have access to best service practices. Specialized method engineers convert knowledge assets from the knowledge base into methods.

It should always be kept in mind, however, that knowledge management programs are not driven by technology, but are enabled by it.

This ends our tour of the ten aspects that comprise the integrated approach to KM at Siemens Business Services.

Perspectives

As a result of the continuous development of the organization of Siemens Business Services, KM as a topic will face more challenges in future. Achieving a certain status is only the basis for motivating employees to attain advanced goals and to continue developing Siemens Business Services into a knowledge-based company.

In addition to this internal view, it is important to position KM as part of the Siemens Business Services portfolio along with topics such as e-business, supply chain management or customer-relationship management etc. Another important factor is the successful incorporation of KM into projects as it was implemented on the basis of knowledgemotion.

Conclusion

There are three basic steps to achieving KM goals. The first step is to determine what kind of knowledge is critical and useful to your business, and how this will best support your strategy. The second is to identify where this knowledge is created, when it is most useful to share it and how this can be done within the context of your organization. Finally, KM processes must be defined as an integral part of business processes. By institutionalizing these KM processes, learning, knowledge creation and knowledge sharing become part of usual, everyday business activity. This means that you no longer have to constantly think about how you and your company should manage knowledge.

Although Siemens Business Services is mainly an IT-focused company, it considers KM as a holistic management approach. Institutionalizing the IT aspects must coincide with possible change management activities. To a large extent your organizational context will determine whether a specific element, for example, an incentive scheme, is necessary to encourage the implementation of KM. One should bear in mind that incentives can only encourage implementation, but will not have an enduring effect on how KM is actually "lived" by employees. Knowledge can only be managed as a natural part of everyday cooperation. This collaboration is typically represented in business processes where individual knowledge-sharing and knowledge-creating processes take place. Rather than creating a KM solution that is implemented on the periphery of the company, Siemens Business Services' approach has been to integrate this demanding task into its daily business activity.

Key propositions

From the implementation of knowledgemotion it was possible to deduce a number of lessons learned for a successful Knowledge management implementation.

1. **Knowledge Management requires problem-based trading.** Knowledge management should not start with an IT solution-based model, but with an in-depth examination of the initial situation in the unit, or the entire company, in order to develop solutions for specific problems (e.g. change management).

2. **Knowledge Management requires support and clear communication of the objectives by management, as well as the active Knowledge Management staff.** This key learning describes the importance of management's role in a successful knowledge management implementation. This cannot be limited to a role as sponsor, but requires an active presence and participation.

3. **To achieve successful Knowledge Management implementation the kind of knowledge that is business critical and its origin or availability must be identified and Knowledge Management defined as an integral part of the business process.** The Siemens Business Services identified project delivery as its core business process, and project experience as one of its most valuable knowledge.

4. **Process owners must be defined and given clearly defined roles and specific responsibilities for output.**

5. **Processes** for the way in which knowledge is captured **must be defined** in the sense of **best practices**, as well as for the achievement and maintenance of the required maturity.

6. **The economic value of knowledge does not lie in possessing it, but in using it.** When the Knowledge Management implementation reaches a certain stage of maturity, having information is no longer the decisive factor for success. It comes down to the manner in which the interpretation and application of this information is utilized. The following comparison makes this obvious: the books in a library obtain their value not by standing on a shelf, but through the readers who read them and use them to increase their knowledge.

7. **The completeness of the Knowledge Management program.** To achieve success, it is necessary to look at and implement knowledge management in its entirety, as in the sections described earlier.

8. **The integration and further development of topic-related projects.** In many groups within a company there are highly knowledge-intensive projects, such as e-business topics, the success of which can be increased by looking at them from a knowledge management point of view. In this context, knowledge management topics should not be run in parallel with such projects, but should be integrated into the projects.

9. **Knowledge management programs must be aligned to corporate goals.** Knowledge management cannot be run as an end in itself, but must be clearly aligned to the strategic objectives of the company in question. At Siemens, for example, these objectives involve supporting the paradigm shift from a product company to a solution and service-driven company.

10. **The provision of a technical platform based on existing architectures.** Knowledge management must not appear simply as a "new" tool to the employees involved; existing information and communication architectures must also be looked at as part of knowledge management project planning.

11. **The pilot projects must have clearly-defined, measurable objectives that can be achieved in less than six months. However, the changeover to a knowledge-based company involves a change process that can span several years.** Planning the pilot projects, in particular, is an important task for the successful implementation of knowledge management. The pilot projects set in motion a process of change, spanning a number of years, which is required to bring about a lasting cultural change. This implies that the relevant employees must understand the benefits of the knowledge management program over and above their intrinsic motivation, even at the pilot-project stage. Furthermore, the pilot groups must be selected in such a way that the results will be significant for business and can be multiplied in other groups or at other locations.

Discussion questions

1. Discuss alternative designs for describing a holistic knowledge management program within one diagram.
2. Discuss the strengths and weaknesses of the knowledge management framework and check for missing elements.

KnowledgeSharing@MED – turning knowledge into business

Dagmar Birk & Manuela Müller

Abstract

This case study demonstrates Siemens Medical Solutions' knowledge management approach, the overall concept and implementation of knowledge management solutions along the value chain, and the integration of these solutions into the processes and the daily workflow. Each solution is described in detail and the potential value thereof is defined for every target group. The increasingly systematic management of knowledge in a proactive division such as Siemens Medical Solutions is explained. Siemens Medical Solutions (MED) will continue to be motivated by its end goals, namely to secure short-term success and long-term viability. A particular knowledge management objective in the support of whichever strategy the division pursues, is to leverage the best available knowledge to make people, and therefore Siemens Medical Solution itself, as effective as possible. An integral part of this case study is illustrating how MED achieves its goal of dealing with operational, customer, supplier, and all other challenges while implementing its strategy in practice.

The corporate context

Siemens Medical Solutions is one of the world's largest and most diversified suppliers of innovative products, services and fully integrated solutions to the healthcare industry. Its offerings extend from innovative technologies for speedy diagnoses to services that optimize processes and increase efficiency in hospitals and doctors' offices. Siemens Medical Solutions has some 28,000 employees worldwide and in the fiscal year 1999/2000 posted sales of 5.1 billion Euro.

Innovation and process improvement are the driving forces behind MED's continued growth in sales and earnings. Inversely the division increases health care efficiency by combining innovative medical engineering with information and communications technology. Two-thirds of our products are less than three years old, and we apply for a new patent almost daily. The growth in the information technology for health care applications is in the double-digit range, while the strategically important takeover of Shared

Medical Systems in Pennsylvania has made Siemens Medical Solutions number one in IT solutions for health care worldwide. The acquisition of Acuson has expanded the range of medical technology systems, resulting in MED now being able to offer complete solutions from a single source. Siemens Medical Solutions is also strengthening its position in the key U.S.A. market.

Parallel to the integration of Shared Medical Systems and Acuson, Siemens Medical Solutions is currently expanding its e-health initiative by selling products and services on the Internet. As the largest application service provider, the Group's e-health includes managing everything from customer data to all processes related to the prevention and treatment of specific illnesses. The goal is to be the world's most successful solutions provider in the health care industry.

A strategic six-pillar approach

The objective for the initiative KnowledgeSharing@MED (KS@MED) was to develop a corporate culture of knowledge sharing – in real time and from person to person – to find swifter, better solutions for our customers. The basic idea is that knowledge created by our employees anywhere in the world should be made available for global reuse. This network of knowledge sharers is supported by a user-friendly, web-based infrastructure that facilitates the capturing and finding of the required knowledge and fosters global joint learning and collaboration. This occurs not only in theory, but on a daily basis, every single work day and across all MED divisions and regions.

The building of the best-in-class sharing capabilities for KS@MED required a holistic and MED-wide approach. To this end the top management of Siemens Medical Solutions had decided on a concept which had been developed during leading knowledge management companies' extensive benchmarking effort. It was then applied and tailored to MED's specific need and currently includes the integration of existing knowledge management initiatives.

Within MED a successful knowledge management approach consists of six key elements (Figure 36) which will be discussed below. A key success factor in implementing such an approach is that all six these elements are well orchestrated and simultaneously refined. Only then can the benefits of knowledge sharing be captured.

1. Top management support

To demonstrate the high priority the topic enjoys and to provide the leverage to overcome barriers, strong and visible support is required from top management. To this end MED nominated Practice Area Leaders. They are top managers with special expertise in one of the following topic areas: leadership, people, strategy, resources, R&D, manufacturing/supply chain, marketing, sales and services, and who are responsible for the steering of knowledge sharing in this topic (Figure 37). The installation of practice leaders for marketing and sales was a primary priority, since Knowledge Management has an extra high impact in these areas.

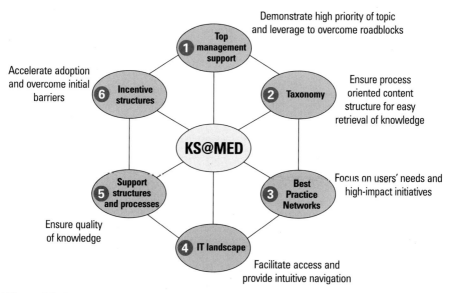

Figure 36 Six-pillar approach of KnowledgeSharing@MED

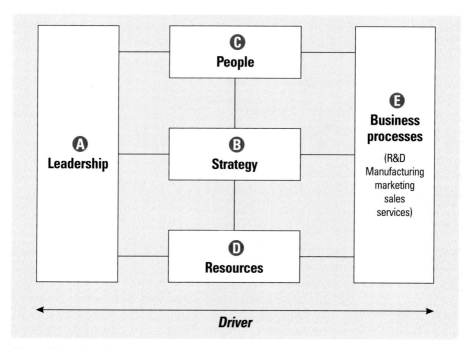

Figure 37 Driver-site of the EFQM model has been used for clear content structuring

2. Taxonomy – define a common topic structure

For easy retrieval of the knowledge contributed, it is necessary to save the information in a common company-wide structure. Siemens Medical Solutions therefore uses a horizontal approach along the value chain, instead of using a vertical structure along the business units. If an employee, for example, is looking for information about logistics, s/he can search in the topic "business processes -> manufacturing/supply chain" to find relevant knowledge from several sources. Without this workflow-oriented content structure it would be necessary to browse through the logistic information of every single business unit.

This horizontal approach is covered by a model developed by the European Foundation for Quality Management (EFQM). The (EFQM) was founded in 1988 by the presidents of 14 major European companies, with the endorsement of the European Commission. The present membership is in excess of 600 organizations, ranging from major multinationals and important national companies to research institutes in prominent European universities.

The EFQM's mission is:

- to stimulate and assist organizations throughout Europe to participate in improvement activities leading ultimately to excellence in customer satisfaction, employee satisfaction, impact on society and business results; and
- to support the managers of European organizations in accelerating the process of making Total Quality Management a decisive factor for achieving global competitive advantage.

Increasingly, organizations in Europe accept that EFQM is a way of managing activities to gain efficiency, effectiveness and competitive advantage. This Ensures longer term success by meeting the needs of their customers, employees, financial and other stakeholders and the community at large.

The EFQM approach provides a clear structure which has been used for content clustering within Siemens Medical Solutions (Figure 37). It is an integral part of every single solution provided by KnowledgeSharing@MED.

3. Identify and support Communities of Practice (CoP)

Knowledge sharing is not only about tools – it's about people. In most large companies many experts work on the same topic. Although they may be situated in different countries. They often don't communicate with one another and sometimes don't even know one another. By bringing the experts together to get to know one another and giving them the opportunity to communicate across borders and time zones, problems are solved faster and better. These groups of experts working on a common topic are called "Communities of Practice (CoPs)". The high impact that knowledge exchange between the experts has on business, lead to the CoPs being one of the first topics to be implemented by KS@MED.

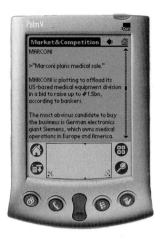

Figure 41
Picture of the Mobile
Business Solution of MED

latest product news as well as information regarding a client's contact person. To this end KS@MED provides a mobile business solution for the sales force. The sales representatives benefits in that they can decide which news they are interested in and download it from a local server on to their organizer.

Our marketing specialists also provide data from the different divisions of Siemens Medical Solutions, generated from different sales systems. The sales systems contain information about products, competitive news, status of the delivery of sold products, lead times for products and the contact data of the different product specialists at the headquarters. With KS@MED's mobile solution the sales representatives receive swift and easy access to the required information, eliminating the need to search for it in different systems (Figure 41).

Balanced Scorecards (BSC)

The Balanced Scorecard is a solution that helps to translate vision and strategy into measurable actions. It enables the executive management to not just focus on capital allocation, but to also emphasize non-financial, critical success factors. Therefore a BSC system reverts to internal data sources and visualizes them in an overview in graphic form. The data displayed in the solution is defined by the BSC and cover the following areas: financial aspects, customer loyalty, business processes and the learning of the organization. These areas are determined by the business strategy of the organization.

Beginning with the goals the organization wants to achieve, it has to be determined what the company has to change to be successful in its strategy. By knowing what to change, it is possible to measure the development of the change – which is done by the Balanced Scorecard.

Key Propositions

1. **Knowledge Management is not only about tools.** People often think of tools by thinking of KM. But KM is more than introducing a tool to the employees. Ensure that all aspects of KM are developed in the solution that you want to introduce to employees. This includes a cultural change with much effort being put into communication.

2. **Align KM solutions to your strategy and integrate them in core processes.** When evaluating what types of KM should be introduced, think about your strategy, and align the solutions to this. A solution-driven strategy, i.e. an approach incorporating efforts from all divisions of the company which use KM, is crucial. Try to integrate all solutions in your core processes, so that KM becomes an integral part of the daily workflow. Focus on this when introducing a solution before you try to introduce a whole range of possible applications.

3. **Content is crucial.** Databases have to be populated before the tool is launched. The first users should find content of high quality in order to be motivated to use the database and to provide content themselves.

4. **Offer individual support.** Most tools used in KM are not really self-explanatory even if the solution is as user-friendly as possible. Individual hotline support is therefore necessary to help users and to motivate them to provide knowledge.

5. **Clearly express the benefits.** In larger companies many databases, content management systems and intranet applications already exist. Clearly explain the differences between these application and the benefits the user can gain from participating in KM.

6. **Top management support.** Do not underestimate the importance of top management's commitment. To successfully introduce KM into an organization you need an expert on KM and top managers who believe in KM. If one of the two is missing, the chance of failure is 60%.

7. **Market success stories.** There should be a continuously survey to determine the success of a solution. Determine what benefits the users gained and then market them in your organization.

Discussion questions

1. Sketch the 6 main elements to introducing knowledge management in an organization and discuss potential additional factors.
2. Discuss the strengths and the weaknesses of the described KM system.
3. Where do you see potential for improvements?
4. State the difference between a KM application and conventional databases or Internet servers.

How to manage company dynamics: An approach for Mergers and Acquisitions Knowledge Exchange

Susanne Kalpers, Klaus Kastin, Karoline Petrikat, Stefan Schoen & Jörg Späth

Abstract

The MAKE case analyses the challenges of mergers and acquisitions (M&A) in an increasingly dynamic company environment and explores methods to improve the success rate. Using existing knowledge management approaches as the point of departure, the application of the approach of Business Communities to the problem area, and how it was adapted for the specific needs of mergers and acquisitions, are critically investigated.

The insights gleaned from the experience described will provide food for thought for others who find themselves caught up in, but ill-equipped to deal with the multi-faceted subject of mergers and acquisitions.

Introduction

Today, when organizational changes are no longer the exception but the rule, the successful management of organizational change has become a key success factor for innovative companies.

At Siemens Information and Communication Mobile (ICM), this insight led to the establishment of a team to critically investigate this challenge and to develop a comprehensive and dynamic knowledge management solution, the Business Community for Mergers & Acquisitions Knowledge Exchange (MAKE).

Mergers and acquisitions at Siemens

Motivation

The only constant is change. This is the message of our times – an era in which change no longer applies merely to product innovation, the deployment of state-of-the-art

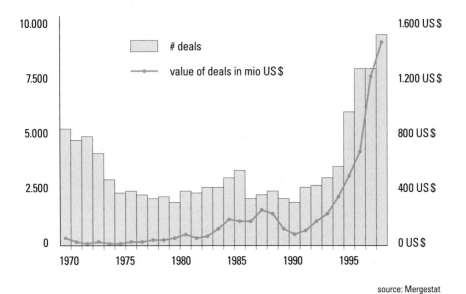

Figure 42 M&A mania worldwide

information and communication technology, and the opening-up of new markets, but increasingly to corporate structures and forms of cooperation as well.

The possibilities for restructuring are many and varied, and range from management buy-outs and outsourcing, to engaging in joint ventures, to the merging of corporations. The motivation for this lies in an effort to achieve market leadership, concentrating on core competencies, swifter adaptation to new markets and operating conditions. The volume of corporate transactions globally has grown exponentially since the 1990s, as can be seen in Figure 42.

Within the Siemens Group, cooperative deals have also taken on a new dynamism. At the shareholders' meeting in February 2000, Heinrich von Pierer characterized Siemens AG as:

...a living organism that must continually change and grow. In fact, change will always be one of the few constants in our company. We can sustain our success over the long term only – and that is the core point of all our portfolio measures – if we attain a leading competitive position in global markets for as many of our businesses as possible. Where we can't manage this on our own, we have the four well-known options: buy, cooperate, sell or close. We are making use of all four options.

However, one important factor in ensuring that a merger-and-acquisition transaction enjoys the desired positive results, is the existence of recognizable synergies, as well as the successful exploitation of the planned synergy effects. Too often financial aspects dominate corporate thinking, while differences of nationality, corporate cultures and working environments, and the incompatibility of IT landscapes, are ignored.

The know-how needed to ensure the successful implementation of mergers and acquisitions plan is frequently only present in the heads of individuals. Siemens' intention is to turn this expertise into common, corporate knowledge and thereby build up a dynamic corporate development.

An introduction to ICM

The Siemens' segment Information and Communications (I and C), with sales of 25 billion Euro in the fiscal year 2000, and about 100,000 employees, comprises three groups:

- *Information and Communications Networks* (ICN) provides end-to-end solutions for voice, data and mobile networks and network components for multi-vendor environments.

- *Siemens Business Services* (SBS) offers solutions and services for electronic business for all levels of the value chain, including consultation and the implementation of I and C solutions.

- *Siemens Information and Communication Mobile* (ICM) designs, manufactures and sells a wide range of communication devices and mobile network products and systems, including mobile, cordless and corded fixed-line telephones and radio base stations, switches for mobile communication networks, mobile Internet solutions, as well as mobile and wire-line intelligent network systems. ICM was formed during a reorganization (that came into effect on 1st April 2000) around the core of Siemens' former Information and Communication Products business group, to concentrate Siemens' activities in the growing mobile communication market. Most of the businesses of Information and Communication Products were combined with the former mobile network business of Information and Communication Networks. As part of the reorganization, Information and Communication Products' former information technology services business was also transferred to SBS. ICM also holds a 50 percent interest in the Fujitsu Siemens Computers joint venture, a leading manufacturer of personal and network computers, and servers. In the fiscal year 1999, the comparable ICM businesses had total sales of approximately 5 billion Euro.

Siemens' mobile phones business, one of the main drivers of its I and C segment, is growing by leaps and bounds. At 11 million units in 1999, sales were almost twice those of 1998. Sales of 25–30 million units are planned for 2000. Siemens expects to market up to 60 million mobile phones in 2001, and aims to have a 10 to 15 percent share of the world market by the end of the 2001 fiscal year.

Mergers and acquisitions at ICM

Prior to the reorganization during which ICM was formed, its predecessor segment Information and Communication Products engaged in a series of transactions intended to focus attention on its core businesses. To this end, from 1st October 1999, Information and Communication Products and Fujitsu combined their personal computer, server and mainframe businesses to form Fujitsu Siemens Computers, a fifty/fifty joint venture, with its headquarters in Amsterdam. Fujitsu Siemens Computers manufactures and markets personal computers, laptops, workstations and servers based on the Windows operating systems, as well as Unix based servers, mainframes and high capacity data storage devices. In the first quarter of the fiscal year 2000, Information and Communication Products sold Siemens Nixdorf Retail and Banking Systems, a manufacturer of point of sale cashier terminals, automated teller and other retail banking machines. At the same time, Information and Communication Products sold its Communication Cables business, manufacturers of fiber optic and other communication cables. To further strengthen its position in the mobile communication handset market, ICM acquired (on 1st May 2000) the mobile telephone business of Robert Bosch GmbH, increasing its engineering capacity for mobile phones operating on the GSM standard.

Mobisphere was another joint venture, initiated in September 1999. Mobisphere is located in the United Kingdom, and was established to share the risks and costs of developing third generation mobile radio infrastructure elements. Siemens holds a 51 percent stake in Mobisphere with its partner, NEC of Japan, holding the balance.

In summary then, ICM's mergers-and-acquisitions projects are as follows:

- September 1999: Joint Venture Siemens NEC – Mobisphere
- October 1999: Joint Venture Fujitsu Siemens (FSC, Fujitsu Siemens Computers), carve-out of Communication Cables and Siemens Nixdorf Retail and Banking Systems
- April 2000: Merger of Information and Communication Products and parts of ICN to form ICM, Transition of ITS from Information and Communication Products to SBS
- May 2000: Acquisition of Robert Bosch GmbH mobile telephone business

Table 1 shows the single process steps for M&A projects at ICM:

Table 1

Analysis and development of co-operations or acquisition/divestment goals and strategies
• Definition of co-operation or acquisition/divestment goals
• Identifying possible partners
• Developing co-operation, acquisition/divestment concepts
• Planning the further steps/procedures
Evaluation of co-operations or acquisition/divestment opportunities
• Evaluation of business strategy
• Screening and evaluation of potential targets

Table 1

Execution of co-operations or acquisition/divestment projects

Project leadership/management for capital investments
Investments, such as acquisition, equity participation, joint ventures
Divestments, spin-offs, joint ventures, outsourcing
including

- the establishment of first contacts

- the management of sounding-out discussions

- the creation of Memoranda of Understanding

- the negotiation of "Terms and Conditions"

- the co-ordination of all Siemens employees involved in the project (or those hired by Siemens)

- the co-ordination of the due diligence

- the development of investment applications for final approval up to contract conclusion/ conclusion of negotiations in agreement with and including the responsible Siemens Corporate departments, management/experts of the relevant Business Division as well as external consultants, if necessary

Approval process for application and approval of capital investments

- Process management for application and approval of investment in the corporate executive committee.

Integration planning and execution

- Developing the integration of the investment/divestment to guarantee a smooth transition of capital investments

- Contract Signing and Closing

- Execution of Integration Plan & Monitoring

Post Closing Management

- Dealing with outstanding rulings on provisions from M&A contracts, as well as providing ongoing support for downstream activities

- Enforcement of contractual claims (Claim Management) plus compliance with contractual pledges

- Management of and participation in subsequent negotiations

- Minimizing of overrun costs and risks of all types (including public liability risks, credit, guarantee and surety risks, plus miscellaneous contract risks)

- Tracking and minimizing of residual costs if the case of sell-offs/demergers, plus keeping tabs on potential synergies in the case of acquisitions/mergers

Legend: | Preparation | | Transaction | | Integration/Post Closing |

General mergers and acquisitions challenges

As illustrated in the figure above, mergers and acquisitions activities go through three phases – preparation, transaction, and post closing (or post merger). Throughout the first phase, the future deal is usually only known to a small circle of top managers and a few people directly involved in the project team. In the preparation phase, the strategic goals of the company, or business unit, are elaborated to define a vision of a future partner. In the time that follows, possible candidates are screened and evaluated for their financial and strategic fit with the company or business unit. After the decision, the negotiations begin. Once the contract has been signed, the first rumors about the transaction start emerging. The closing becomes effective with the legal transfer of ownership.

Until the closure, the activities take place at a high level and very few people are involved. The most relevant phase, however, the one that finally determines success or failure, is the post-closing phase where the integration of the carve-out of the business takes place. In this phase, the transaction becomes public, many people are involved, and the hoped for synergetic effects have to be realized.

The potential synergies are enormous, the analysts' promises in advance of a merger are bountiful – and yet the desired effects often fail to materialize. Consultants have learned that:

"83 percent of all mergers fail to improve shareholder value". (KPMG)
"85 percent of all mergers fail to meet the objectives they set out to achieve".
(A.T. Kearney)

This is hardly surprising. In most mergers and acquisitions transactions the individuals charged with implementing the plans are performing this function for the first time. There is a shortage of contacts and information, and the scope of in-house expertise and decision-making is not clearly defined. Integration tasks – not that they are easy to define at any time – tend to get buried in the day-to-day business. After all, everyone knows that "keeping the business going" is of primary importance. This means that checklists are prepared, rules drawn up and de facto solutions devised "on a wing and a prayer", with the overall interrelationships taking a back seat. The top management, who originally formulated the contractual conditions for the transaction, has long since moved on to other tasks.

In order to counter possible negative effects of mergers and acquisitions transactions and to ward off or minimize risks, Siemens has established a "Post Closing Management" (PCM) department at corporate level. This department's primary function is to ensure that the requirements of third parties arising out of mergers and acquisitions transactions are consistently met.

General deficits of mergers and acquisitions activities

From studies of mergers and acquisitions activities and the experiences described in literature, it is possible to conclude that although mergers and acquisitions have become increasingly important, they are, in general, not yet seen as separate business processes. This applies in particular to the integration and consequent realization of synergetic effects. While the strategic and contractual parts of the deal are already sufficiently taken care of, the integration, which finally involves the whole organization, is not yet seen as integral part of the whole process.

Mergers and acquisitions differ greatly in size, scope, and so on, which makes standardizing the characteristics difficult. However, there are certain general factors that are found in successful mergers and acquisitions and, consequently, it is possible to postulate some reasons for the failure of mergers and acquisitions projects as well:

Communication/buy in

Every mergers and acquisitions transaction leads to changes. Initially these changes are not known. Once the merger is announced, people begin to wonder if their jobs are secure, how structures will differ and what their roles in the post merger integration may be. They are often disoriented and skeptical about the outcome. The only way to assuage these concerns is to share the vision and strategy of the merger and the steps that will follow with the employees from the beginning. Communication throughout the merger, from the first announcement to the end of the post-closing phase, is crucial for a successful merger.

Vision/strategy

In many cases, the predefined processes end with the deal being closed. The people who are in charge of the integration are not the ones who have signed the contracts, and are consequently left without information about the vision and strategy of the merger. Ideally, the "integration managers" should be phased-in during an early transaction phase. Similarly, the visions and strategy for the post-merger integration should be developed before the contracts are signed. It is important that the post-merger integration is seen as an integral and vital component of the entire mergers and acquisitions process.

Integration team

If integration is left unplanned, nothing gets done. If everybody is called to integrate, without being assigned concrete tasks, it is unlikely that it will ever happen. The integration has to be lead by a core project team with a definite project plan, milestones and – most importantly – enough capacity to reach the goals. In a fast changing environment with vital decisions being made, a core team should co-ordinate the integration activities. If integration is seen as a part-time job, or something that will take care of itself, its success is endangered.

Experience/information

Though mergers and acquisitions are beginning to become an integral part of business life, each deal is different and many of the people involved are novices at their jobs.

Consequently, there is a need to build or collect information or knowledge that is common to all mergers and acquisitions and their successful management. Experience reports, guidelines, circulars, and so on can prove to be incredibly helpful, but are hard to find. A collection of experienced contacts could be the start of sharing knowledge and information about mergers and acquisitions.

Project MAKE reflects the dynamics

The beginnings of MAKE

A MAKE team member remembers:

The realization that we had to support our mergers and acquisitions projects more effectively, became clear last year. Everybody was suddenly involved with the carve outs of Fujitsu Siemens and Flextronics. The subject dominated conversations because all aspects of our work, as a Chief Information Office, were involved. Just a few days before the due date, a colleague of mine learned that all the email addresses had to be changed. Some other person I know was told to evaluate a Swedish company's information and communication infrastructure in the due diligence phase of a merger activity. At that time he didn't even know what "due diligence" meant.

The motivation behind the founding of a new department, ICP Post Closing Management, in October 1999, was this awareness – that mergers and acquisitions projects were not over and done with after the final signing of the contracts. Simultaneously, the subject gained topicality by showing up on management slides. The Strategy & Business Alliances department, that played a leading role in most of the mergers and acquisitions transactions by providing the related expertise and project management, created a mergers and acquisitions reference book. This book provided summaries and guidelines of their experiences in that field to support the people who were working on mergers and acquisitions deals at that time. This book also described the previously mentioned steps of the mergers and acquisitions process. The content is now available both as a binder and electronically for the downloading of checklists, forms, standard contracts and lessons learned reports extracted from various mergers and acquisitions projects. This knowledge base reflecting the overall mergers and acquisitions process and covering experiences of several mergers and acquisitions deals, became not only a working tool for people involved in mergers and acquisitions projects, but also a building block of the MAKE project.

A consultant from the corporate technology department remembers:

"When we entered the project, its scope was quite vague. The goals ranged from defining the mergers and acquisitions process for IT aspects, to knowledge support for the ongoing mergers and acquisitions projects. As experts in Knowledge Management, we came along with the idea of building a process for knowledge sharing and creation, instead of just providing static information. In January everybody at Information and

Communication Products (now ICM) was talking about future mergers as a general topic. They probably already knew by then they would be merged internally to build the IC Mobile group.

The project was originally supposed to focus only on the IT aspects of mergers and acquisitions transactions, but in the initial workshop it became quite clear that the IT aspects were deeply interwoven with the other aspects of the mergers and acquisitions process, and that this affected nearly all functions and aspects of the corporation. By March 2000, the team presented the concept and the first prototype information platform for the newly borne MAKE initiative – the Mergers and Acquisitions Knowledge Exchange.

Initial team meetings

Though the project is still in its infancy, the personnel changes have been significant. Two members of the original core team left, after contributing significant information and documentation. Most of the changes resulted from the reorganization that culminated in the formation of ICM, the new Siemens group. The project management was subsequently given to a new member of the department, Jörg Späth, who brought fresh perspectives from his own experience in another Siemens group. The newly enlarged Strategy and Knowledge Management department appointed a new head and last, but not least, a new Chief Information Officer was appointed at ICM.

The changes were no surprise as a merger brings changes: managerial changes, cultural and strategic changes. The changes at ICM only reflected what the project actually deals with: the increasing dynamics within a company and how they can be managed.

The staffing of the project team was initially not quite clear, as the experts were very busy with priority mergers and acquisitions activities. Nevertheless, the participants of the first workshop decided to start building a Mergers and Acquisitions Community.

The highlights of this first workshop were:

- The discussion about the process model for the mergers and acquisitions integration phase.
- The general insight that communication problems were the main pitfalls in the early stages of a merger. An "information broker", who could keep people posted about decisions, personnel changes, circulars etc., was therefore proposed for each mergers and acquisitions project.
- The creation of the term "knowledge package", as a unit of coherent information about mergers and acquisitions.
- A discussion about the classification criteria for knowledge packages (see: M&A Content Spider).
- A detailed list of the steps to be followed (in the project).
- A list of people who could be contacted.

The goal of the second workshop was to integrate new members into the team and to present and discuss the elaboration of the classification criteria for knowledge packages. The event was graced by the presence of two people with a great deal of mergers and acquisitions experience:

- Gregor – who is responsible for the integration of organization and information services of the newly merged departments ICP CD and ICN CA at ICM – was first introduced to the project team. He was a little reluctant to join the team for two reasons: 1) He was not sure he would find the time and did not know if the MAKE project could be helpful in his current position. 2) He was not sure if he could contribute anything of value, because he felt he had himself not yet learnt enough about mergers and acquisitions. Fortunately, he was soon convinced that he had valuable experiences to contribute, when he learned that there were other experts who had had to deal with similar problems. At last he had met people with whom he could discuss his concerns.

- Peter – an employee of the Corporate Information and Knowledge Management Department, responsible for the IT infrastructure issues of all carve-out, merger and acquisition activities at Siemens – was very excited about the MAKE project. As a knowledge management project, it was exactly what he needed to enhance his work. He is very aware of the "nuts and bolts" of the integration, as the most common crucial problems in the early phase concern the IT infrastructure. He says, *"We cannot wait until the fog vanishes and the management decides on a strategy. Since we support a real business process with real infrastructure, we need to act here and now"*.

Peter actively advocates more transparency in the mergers and acquisitions integration phase. He envisages a common strategy that can be reused for all mergers and acquisitions projects. He often takes on the role of an information provider for an entire mergers and acquisitions project and is aware that there is no clear line between IT infrastructure and other mergers and acquisitions topics, like culture or contracting. He facilitates the smooth progress of the project by contributing his practice-oriented ideas, anecdotes and success stories.

In the third workshop, the prototype of an Intranet information platform for knowledge exchange was presented, together with a strategy to make it viable – this will be discussed in more detail later.

Though it is obvious that a technical solution is not the key factor in the final success of the project, the effect of this Intranet information platform on the workshop participants was impressive. The project members' ideas were now all moving in the same direction. The manifestation of MAKE became visible and tangible, and formed a good basis for discussions. The team decided to start using the platform immediately to exchange knowledge about its project results. Moreover, the scope was broadened to encompass all central functions (departments) and all mergers and acquisitions aspects. Expectations were growing because the solution contained possibilities that had, to date, not been at all clear to most of the participants. Along with these expectations came the hope that the team was rapidly achieving its goals – which are described in the next section.

The MAKE approach

The main goal of the MAKE project is to implement a knowledge management solution that will improve mergers and acquisitions activities at Siemens ICM. The project has the following objectives: to reduce recognized deficits and have a positive impact on the efficiency, the speed and the success rate of future mergers and acquisitions. Consequently, the anticipated benefits are substantial.

The necessity for Knowledge Management in the field of mergers and acquisitions stems from the intention to collect and share knowledge about mergers and acquisitions processes and thereby multiply experiences. The more efficient handling of knowledge in this field should lead to the realization of synergies and practical use of the knowledge, and should also avoid knowledge loss.

Knowledge management solution

There are different knowledge management strategies with which to tackle the challenge. From the wide selection of knowledge management approaches that are of practical relevance (for example, training, Yellow Pages, knowledge mapping, expert systems, business intelligence services, lessons learned process, competence center, etc.), the team considered the idea of Business Communities (i.e. Communities of Practice) as the most promising here.

A Business Community is a group of people who share existing knowledge, create new knowledge and help one another on the basis of a common interest in a business-relevant topic. The group is usually distributed by geographical location and organization. The relevant people pursue both individual and business goals through both face-to-face and virtual collaboration.

The knowledge required to successfully implement mergers and acquisitions is highly complex and case-dependant. The important experiences and know-how are not suited for capture in a database. However, the knowledge is of critical future relevance for the business, and the company-specific part, in particular, cannot be acquired from external sources.

As mentioned, the people who have mergers and acquisitions knowledge are distributed over many different organizational units, locations, business processes and projects. It is therefore crucial to build a network of all relevant people in the different mergers and acquisitions activities: people from central functions who are involved continuously, or repeatedly, in mergers and acquisitions, as well as the people who are involved only once when their organizational unit is involved in a merger or a carve-out. Interest in knowledge exchange may well be temporary for many of those involved, especially if they have been or are only involved in a single merger. In carve-out situations they may even end up no longer being company employees. For obvious reasons, these people are from all kinds of disciplines (e.g. Information Technology or Human Resources), and therefore do not necessarily use a common terminology, which poses an additional challenge.

The Business Community is designed as a socio-technical system with the objective of implementing the overall process consisting of

- the individual documentation of knowledge
- the contribution of information to a shared information platform
- the quality assurance and condensation of the collected information over time
- the active communication of knowledge to others in face-to-face and virtual interactions and the brokering of knowledge
- the searching for, and access to, available knowledge sources (information and persons)
- the combination of knowledge and the creation of new knowledge
- the use of knowledge and learning.

The MAKE approach represents the team's effort to try and find an appropriate mix between the documentation of existing knowledge and the networking of relevant persons for direct communication and collaboration.

The overall knowledge management solution, which has to date only been partially implemented, consists of the following five components which will be described in more detail:

- The Business Community concept
- The information platform
- Knowledge packages
- The mergers and acquisitions process model
- Mergers and acquisitions project coaching

The Business Community concept

Any knowledge exchange starts with the people who participate in it. People, therefore, are the main focus of MAKE. The vision is to create a dynamic group of experts who are linked by a network of communication streams and who use a common language about mergers and acquisitions. Everybody in this group contributes and has access to a variety of material about past and current mergers and acquisitions and can easily search for a specific topic. Everybody knows whom to contact with an urgent question, or knows somebody who knows. Experienced, well-known mergers and acquisitions professionals are drawn in early on in new mergers and acquisitions projects. The group develops a common process, adapts it when changes are required, and exchanges experiences and anecdotes.

The development of the MAKE community went through different phases: the awareness phase, the start-up phase (both now completed) and it is now in the activity phase. During the awareness phase, the team identified the actual deficits, expectations and potential areas of improvement of Knowledge Management. Starting with the first meeting, it developed a core group that forms the MAKE project team and the heart of the business community as well. The core group is a carefully chosen combination of experts in different aspects of mergers and acquisitions.

In the second phase, the start-up phase, the team analysed the current situation, using a systematic knowledge-management assessment tool, and developed a concept for the implementation of the desired knowledge management solution – the Business Community MAKE.

With each following workshop, the team established not only the MAKE strategy, but also personal relationships and a common terminology.

The team used the developed Business Community concept to inform and motivate other prospective members about the MAKE approach and invited them to join the group. The real Community life began with its kick-off workshop. The MAKE concept was discussed and adapted to the group's needs and expectations. The number of members in the Business Community is currently growing. The challenge now is to sustain the high motivation of the start-up phase and to achieve the expected positive impact.

The common concern of the Community members is to reuse experiences, to have a good process model with checklists, methods and tool, to have access to relevant information and valuable contacts, and thus provide solutions to their day-to-day working problems. The Community tries to meet these needs in various ways, making use of both planned and regular communication, as well as more continuous and spontaneous communication. To ensure the right environment, the main activities receive some support. This is primarily the job of the Community moderator and the Community facilitation. The table 2 lists the various activities the Community engages in.

Table 2

Main activities of the Business Community:	Support activities for the Business Community:
– Sharing regular events (face-to-face and phone conference) – Urgent request forum (Discussion forum with e-mail notification, in addition to phone calls and Net-meeting sessions) – Information-platform process for knowledge packages, contacts, process models and project information (contributing information, assuring quality, condensing and linking information, accessing and using information) – M&A process improvement workshops – Coaching and accompanying of current M&A projects: information brokering and project debriefing	– Topic and activity management – General CoP organization (financing, group structures, etc.) – Member management – CoP external promotion (e.g. transparency in corporate community landscape) – Development of relationships with sponsors, management and related groups – Support of the group-identity – Provision of guidelines, templates and checklists for the main activities – Definition of content structures, improvement of shared context and common terminology – Provision and administration of useful information & communication platform – Collection of success stories, continuous improvement of the knowledge management environment

The MAKE Business Community is intended to be internal to the company. Most members come from the Siemens group ICM, some experts are from corporate departments and there are links to other relevant activities within Siemens. At the moment, the Community is relatively small with fairly active members. In future, three circles of involvement are planned:

- *the core group,*
  the moderator and one owner for each mergers and acquisitions aspect (sub-topic area)
- *the inner circle,*
  the active long-term members (for example, who attend the regular sharing events)
- *the outer circle,*
  temporary or less active members.

The role of the moderator is officially recognized and financed by the company.

The MAKE community is visible within ICM and throughout Siemens, and integrated into current and future mergers and acquisitions activities. The membership of the group is restricted and membership is only possible through the core group members. The information platform is partly open within the Siemens Intranet. There is no special incentive system – the motivation to contribute to and be active in the Community derives from the benefits to individuals in their daily work.

In the existing MAKE Community the following roles are operative:

- *the Community members,*
  who contribute and use knowledge by participating in Community events or workshops, joining the urgent-request process and using the information platform
- *the MAKE moderator,*
  who is responsible for the Community support activities and the overall moderation of the main activities (see table above)
- *the topic owners,*
  who are experts on certain sub-topics
- *the information platform editorial team,*
  which is responsible for the content of the information platform, and provides passwords, etc. It comprises a MAKE moderator, an IT administrator, all topic owners and content assistants
- *the mergers and acquisitions project leader or the information broker*
  who is responsible for the information about a specific mergers and acquisitions project and motivates his or her project team to participate in the MAKE Community.

The information platform

The Business Community MAKE has its virtual home in an Intranet-based interactive Web application (see Figure 43). The site is partly access-protected and its content maintained interactively by the Community members. Major changes are discussed in an editorial team.

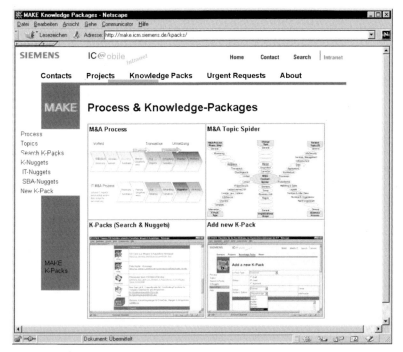

Figure 43 The MAKE information platform

This application reflects the insight that a heterogeneous Community has a wide range of demands and, eventually and ideally provision ought to be made for each type of knowledge involved in mergers and acquisitions, however structured, as well as for multiple ways of communication.

The central idea is to classify each piece of information, from newspaper articles to telephone numbers or anecdotes, by attributing it to certain categories and then storing it as a "knowledge package" (K-Pack – see later explanation). Basically anything on the website is a knowledge package, but the main types of K-Packs are accessible in the primary navigation of the platform.

- In the *contacts section* one can find the co-ordinates of the entire Community, along with the specification of their expertise, the project they are currently working on and their role in the Community. Contacts to external experts and organizations are also displayed here.

- The *projects section* provides the user with the most important information and links to further sources on running and completing mergers and acquisitions projects. Data for new projects, or changing data of current projects can be done interactively.

- The *knowledge package section* contains data on the mergers and acquisitions process and the categorization of K-Packs. A sophisticated search engine allows the user to keyword-search the K-Packs. It is also possible to browse through the most popular and often used K-Packs, which are called "K-Nuggets".

- The *urgent-request section* is a discussion forum where Community members post their questions to the Community and discuss current issues. These electronic discussions are supplemented by direct phone calls or Netmeeting sessions. Answered questions and solved problems are archived and, in special cases, added to the FAQ list.

- In the *news section* the moderator and the Community members are kept up-to-date about activities and events.

- The *about section* provides the user with information on how to utilize the IT platform and the roadmap of the MAKE project. The different roles and activities in the MAKE Community are also described.

Knowledge packages

The problem areas of mergers and acquisitions are as numerous as the many different types of organizational changes: from intra-company mergers to the carvings out of small technology start-ups. To facilitate the search for reusable knowledge and to structure the knowledge, the team has developed a "spider" in which each leg represents a dimension or category (see Figure 44). The categories then serve to classify the knowledge packages.

The scope of a knowledge package is as wide as possible. It can be filled with any kind of information about mergers and acquisitions activities, without any formal restric-

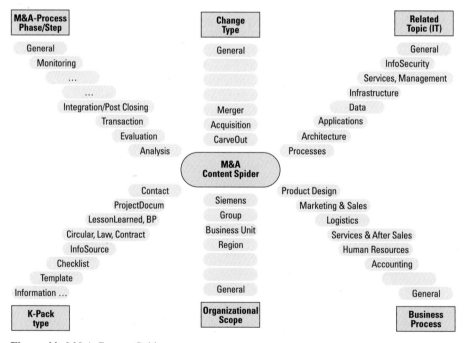

Figure 44 M&A Content Spider

tions. This allows the system to adapt to the kind of information that is most useful to the Community.

Every knowledge package is added to the platform with its attributing meta-information. The meta-information serves to give an overview of the K-Pack's content, the author, and its current status and to allow a better keyword search. Much of the meta-information is optional. An example is given in the table below.

Table 3 MAKE Knowledge Package (K-Pack)

Meta-Information	Example
Header Information	
Title	Checklist for the execution of organizational changes
K-Pack Type	Checklist
Status	Approved
Owner/Author, Editors	Owner: H.H:, ICM CIO CF Editor: S.K., Project MAKE
Description	
Spider Categories	Organizational Scope: Business Unit Change Type: General Business Process: General
Abstract	The checklist provides an overview of items that have to be considered during the execution of an organizational change. Each item has to be checked for its relevance, additional items can be added for individual cases.
Keywords	Checklist, organization, unit, process, internal, change
K-Elements	Word — Checklist_ICP_19991030.doc
	HTML — http://...../siemens.de/.../orgweg/orgaenderungen.htm
Search	Search for related K-Packs
General Information	
Lifecycle/Dates	Creation: 1999 / 10 / 30 Last modified: 2000 / 04 / 07 Check again: 2001 / 04 / 06
Access/Security Level	Internal
K-Pack Is	Sc991030a

Initially, the core group collected interesting and reusable information and made it available in the form of K-Packs. The platform also allows all Community members to add interactively to their knowledge packages, which are potentially large and grow fairly continuously. The topic owners are responsible for monitoring and improving the usability of the K-Pack within their respective topic. From time to time, the K-Packs have to be reviewed for relevance by their owners. In special cases, correlating and similar K-Packs are condensed to a shorter new one, which provides a good overview and summarizes the most important points. Moreover, often-used K-Packs are listed for easy access in the K-Nugget section.

Improvements to the mergers and acquisitions process model

The Community wants to improve the existing process model based on the following insights, among others:

- The integration/post-closing phase is crucial for the success of a mergers and acquisitions transaction and needs to be detailed and embedded.
- The communication about the changes in key roles, restructuring, etc., should be an essential component of the process model, because uncertainty and confusion are contra-productive.

In the MAKE Community, the experts on the various aspects contribute to developing and activating the mergers and acquisitions process.

Mergers and acquisitions projects coaching

Information broker

As was stated earlier, communication and buy-in factors are significant factors for the success of a mergers and acquisitions transaction. It is a long communication line from the first top-secret negotiations to the public announcement and buy-in of all employees and customers. The levels of secrecy at the various project stages should be described in a communication concept as part of the project plan.

When the merger is announced, the employees have to be included in a clear strategy and vision.

In the post-closing phase, the decisions about the changes concerning leadership, guidelines, organizational structures, and so on, have to be communicated as soon as they are made. Since the post closing integration team is busy enough getting things done, the team is supported by an information broker, who has the following responsibilities:

- to post all information about recent changes and decisions on the platform
- to provide useful information: guidelines, circulars and laws
- to communicate the progress of the post-closing project
- to help people find their contacts in the merged organization
- to be the main contact person for the discussion of merger issues.

The information broker is an active member of the MAKE Community and uses the information platform as a communication channel for his specific mergers and acquisitions project.

Project debriefing

Once a project has been completed, there is usually too little time to contemplate the learned lessons. In the case of mergers and acquisitions, these experiences can be extremely helpful to others in the company, so a formalized debriefing is an integral part of every mergers and acquisitions project. The MAKE core team coaches and supports this activity. Many lessons, useful information and reusable methods or tools from all the completed mergers and acquisitions projects, will be made available to the MAKE Business Community in future.

Conclusion

The application of Knowledge Management to the field of mergers and acquisitions is both promising and challenging. The scientific value for the knowledge management field lies in the expansion of the networking concept (e.g. Business Communities) to a field where some of the usual prerequisites do not apply. The methodology of Business Communities must be developed further and adapted to the specific needs of the mergers and acquisitions context.

The MAKE project has already had a considerable impact on current mergers and acquisitions projects, e.g. the reusability of experience and a progress in people's awareness of the mergers and acquisitions domain. However, the final success can only be evaluated once the Business Community is fully functional and self-sustaining, and the components of the developed concept have been fully implemented. It is important that at a defined future point, the applicability of the project's approach to domains with similar features and shortcomings is further investigated.

Key propositions

1. Mergers and acquisitions is a major challenge for innovative companies in a rapidly changing environment. The number and volume of corporate transactions such as mergers and acquisitions are increasing drastically. Companies are driven by the expectation of positive merger results. However, the exploitation of the planned effects is only possible when organizational changes are actively addressed as a challenge.

2. The development and communication of a well-defined mergers and acquisitions process is important. Investigations about the deficits of mergers and acquisitions activities have shown that in many cases the organizational change is seen as a one-off event and not as an instance of a generic process. In order to learn from experience, it is crucial to use and communicate a well-defined mergers and acquisitions process that can be adapted and improved over time.

---

3. Business Communities are a useful knowledge management concept for mergers and acquisitions. The Business Community approach is applicable to the field of mergers and acquisitions, because the knowledge is highly complex and can only be collected, shared and reused through a network of experts. The people involved in mergers and acquisitions share an interest in a problem domain of high relevance to economic success. However, they are distributed by location and organization.

4. Integrating Knowledge Management in the mergers and acquisitions process means both
   • improving the information flow during the project and
   • making experiences from other mergers and acquisitions projects reusable for all relevant persons.

5. Experience shows that in mergers and acquisitions projects communication about the status of activities, contact persons and current decisions is of outstanding importance. Knowledge Management can leverage the communication flow by providing a frame and a platform for people to post and exchange all relevant information about the project. Through comprehensive project debriefing it becomes possible to store and distribute useful experiences to all relevant persons.

Discussion Questions

1. Give a rough sketch of the mergers and acquisitions process and reasons why mergers and acquisitions often fail (do not reach the expected results).

2. State reasons why Knowledge Management for mergers and acquisitions is important for companies like Siemens.

3. Describe the main requirements for a knowledge management solution in the given situation.

4. Discuss the MAKE knowledge management approach. What risks and barriers do you see endangering a successful and sustainable implementation of the process? What alternative approaches or modifications would you regard as useful?

V   Learning and Knowledge Management

Knowledge Master – a collaborative learning program for Knowledge Management

Christine Erlach, Irmgard Hausmann, Heinz Mandl & Uwe Trillitzsch

Abstract

Like every new business topic, Knowledge Management is a new learning area for employees engaged in this issue. The difference with Knowledge Management is that it is complex and multi-facetted. It requires not only knowledge about methods or tools but also a certain state of mind and creative competence. So classroom teaching, reading, or simple tool-training are not ideal solutions to this problem. Approaches that are more integrated are needed for learning about Knowledge Management.

This case study illustrates the need for and benefits of a corporate learning program for Knowledge Management and its implementation. The realization of the program, called Knowledge Master, through a public-private partnership is described, as are the program design and the virtual learning platform. Participants' experience of the program is also reviewed. The study closes with a look at critical issues in the design of successful corporate learning programs on Knowledge Management.

A Knowledge Management conference

At an international Knowledge Management (KM) conference, Irmgard Hausmann, Siemens Qualification and Training's (SQT) project manager of Knowledge Master, is finishing her work on the exhibition booth of this further education program, before the doors of the conference hall open and the participants stream out to visit the various booths.

"Hello, Mrs. Hausmann".

Irmgard turns around and recognizes Peter Müller, a participant in the pilot program of the Knowledge Master course. "Are you attending the conference, Peter?"

He laughingly replies: "No, I'm giving a presentation on our experiences with the implementation of Knowledge Management".

"That's very interesting. It doesn't seem as if you needed to attend the Knowledge Master course last year. Why did you?"

"That's quite a story! I am an IT expert who was suddenly transferred to our knowledge management project. I didn't have a clue what Knowledge Management was all about and didn't have the time to study it either. Then I heard about Knowledge Master and it seemed to be the answer to my problems. Learning and practicing without having to take time off from my job sounded very attractive indeed. Doing the Knowledge Master course was exactly the right thing to do as I was promoted to project leader of our knowledge management initiative. I am now even able to address this conference as if I have always known about Knowledge Management".

Irmgard smiles, vividly remembering the trouble she had in finding enough participants for the pilot course. People had been so skeptical about the concept and its chances of success. Her enthusiasm and that of the University of Munich partners had sustained the project. They had all seen the need for a practice-oriented training program for Knowledge Management – a complex subject if ever there was one.

She is curious to know more about Peter's experiences during and after the Knowledge Master course. They decide to get together again during the next coffee break to continue their conversation.

Pensively Irmgard thinks back to the start of the development that eventually led to the Knowledge Master course.

The development of a learning area

During the early 1990s, knowledge increasingly became the basis of the business of the electronics industry. The growing recognition of knowledge as a productive factor, led to the first trials to better manage this asset. The term Knowledge Management may not have been used as yet, but this was what was being done. Siemens was a pioneer of this development.

Knowledge as a productive factor required a new way of thinking about business and management processes, and strategy and business models, but the means to build competence in these fields did not seem to be available.

As a solution to this problem, the knowledge management pioneers founded a Community of Practice (CoP) to share their experiences and to discuss contemporary knowledge management issues. When Knowledge Management became an accepted term and a company-wide issue at Siemens in 1996, this CoP was the basis for the Corporate Knowledge Management (CKM). CoP CKM was made official in an effort to bring together all available competence in this field. It comprised about 30 people from business units, corporate functions, and internal services. Representatives from Siemens Qualification and Training (SQT) were also among the founding members of the CoP CKM.

Over the next year, the CoP CKM expanded. Many who were new to Knowledge Management joined the community, as knowledge management tasks were increasingly being required of employees. As a result, the question was raised in the CoP CKM of how competence in Knowledge Management could be built, besides simply learning through experience. Despite the plethora of conferences, seminars, studies and White Papers on Knowledge Management, the requirements of beginners in this field were not being met. The material offered was often too superficial or too narrowly focused to equip a future knowledge manager, whilst the number of jobs requiring knowledge management tasks continued to grow.

A good further-education program for this subject was urgently required.

The requirements of a learning program

A gap in the market, the unsatisfying alternatives being offered by competitors and a growing demand for managers with knowledge management skills, gave Siemens the motivation to embark on its own product development project. Siemens Qualification and Training (SQT) decided to take action.

SQT is ideally positioned to initiate a learning program like this one. As the Center of Competence for consultation, operational support, and training within the Siemens group and an international full-service provider in the qualification business, SQT has had vast experience in the further education market. Its high level of competence in finding business solutions and in organizational and personnel development, and its comprehensive business-related know-how, allows SQT to offer its customers tailor-made solutions to their qualification problems. Their expertise ranges from international business skills to multi-media applications, to managerial competence and knowledge of business methods.

Initially, there were questions for which SQT sought to find answers:

- What does a good knowledge management training program look like?
- What are the basic requirements of such a program?
- What are the criteria of an excellent learning program, in general?
- What can be learned from other such programs?

As Siemens' internal market would be the main target for the training, SQT already had customer confidence. Its profile, as the preferred training supplier to all Siemens employees, was an added bonus. Specific knowledge of customer needs was readily available from the CKM Community.

The corporate learning environment versus the academic learning environment

To begin with, SQT determined that their corporate learning program had to differ from learning programs in academic settings. This implied:

- It had to not only provide scientific knowledge ("Why" knowledge), but also practical knowledge drawn from the experiences of other companies ("How" knowledge).
- It had to be able to be pursued part-time, parallel to full-time employment.
- It had to integrate case study material and the participant's own business problems.
- It had to combine classroom teaching and workshop-style learning.
- It had to stimulate the participants' critical and conceptual thinking.
- It had to be based on a sound didactic design that fostered the transfer of the contents learned to everyday life.

In turn, the contents should be adapted to the corporate environment. The contents should

- provide the learner with a state-of-the-art overview of knowledge management philosophy, as well as the tools and implementation process of a knowledge management program
- include the full range of the different knowledge management activities (e.g. knowledge acquisition, creation, sharing and utilization)
- consider the specificity of individual Knowledge Management, as well as that within a team, or an organization as a whole
- take into consideration the different perspectives on Knowledge Management (social, business, and technical) and the related knowledge management pillars, namely human, organizational and information technology
- enable the learner to gain experiential knowledge about knowledge management processes
- simulate the real knowledge management environment, for example, by asking the learners to work in virtual teams on authentic cases.

With the goal of developing an outstanding program, SQT's challenge was to integrate the newest learning methods and pedagogical insights. From this point of view the program had to

- utilize a problem-based learning approach that focused on learning rather than on teaching
- recognize the social dimension of learning
- enable learners to acquire practical competence in order to take effective action
- pragmatically combine substantial depth, practicability, and applicability
- simulate the reality of Knowledge Management by including several disciplines to give it an interdisciplinary aspect

- take modern pedagogical principles based on constructivistic approaches into consideration.

It soon became clear that such an ambitious training program required diverse competencies; this vision would not become a realization without the participation of partners.

A collaborative project

The core competence of SQT is geared to setting up and running training programs. While SQT anticipates future business needs, defines business opportunities, develops training concepts and takes care of the quality of its training programs, the detailed development of the training material and the actual presentation of the training are done together with external partners. This eliminates the need for SQT to provide up-to-date content on every topic offered. The Knowledge Management project was no exception, and SQT began the search for suitable partners.

SQT had worked with various departments of the University of Munich on other projects in the past, so it was an obvious choice as a development partner. They understood the concepts and had experience of knowledge sharing, creation and didactics. The initial meetings between the relevant university staff and SQT proved fruitful: both parties were interested in a collaborative effort.

Although a university is often said to be in a world of its own, which differs from the realities outside its walls, one of its goals is to equip its students to accept responsibilities in corporations and society. This project came as a welcome opportunity for academics to work on a "real" project in the "real" world. They were willing to collaborate with Siemens if the university could provide the contents and didactics of the training program, and if Siemens could provide real-life examples and were willing to extend the training to the university's students as well. It was envisaged that students from the university and employees of Siemens and other companies would then be able to learn together, each sector contributing its unique competencies from the vantage point of their different perspectives on Knowledge Management. After a series of consultations, the concept of the training program Knowledge Master (the name of the new program), emerged as a public-private partnership between SQT and the University of Munich.

A further education program on Knowledge Management required a complex pedagogical design. Prof. Heinz Mandl, Director of the Institute of Educational Psychology, provided the initial input. His team developed the didactical design of Knowledge Master. The Institutes of Management and Information Technology, headed by Prof. Picot and Prof. Wirsing, respectively, then enriched the concept with their expertise and provided state-of-the-art content for the training.

Knowledge Master's goals and concept

The goal of Knowledge Master is to enable participants to manage and initiate knowledge management projects in their working environment. Each Knowledge Master graduate should know what Knowledge Management involves and what benefits it provides. They should be able to formulate concrete goals and strategies that suit the business context and realize them. Each graduate should emerge educated, competent, and motivated to implement Knowledge Management.

This complex goal necessitated choosing a didactic concept that would support the development of applicable knowledge. Modern educational theory asserts that the transfer of learned contents is best fostered by the modern constructivistic approaches of self-guided and collaborative learning, which rely heavily on the cognitive apprenticeship structure. This led to the formulation of a concept of learning from material drawn from authentic cases, within a social learning environment.

Knowledge Master aims to provide an integral understanding of Knowledge Management. To this end, the social, business and technical perspectives of Knowledge Management are presented by the Institutes of Educational Psychology, Management, and Information Technology, respectively. In this way, the participants learn to appreciate the different perspectives on Knowledge Management and achieve a more integrated understanding of it.

As collaborative management training should also ideally provide opportunities to develop practical competence in the field of Knowledge Management on a personal, team, and organizational level, Knowledge Master combines classroom teaching, workshop sessions, self-guided learning, and virtual learning. In this way, the design of the program facilitates the linking of the learned content with the demands of daily business and the personal experience of the participant. Through the social learning environment and the virtual learning platform, the participants build a Community or group of learners, with everyone contributing their personal strengths and experiences. The group works on team assignments, which, in turn, reinforce the building of knowledge management structures within the group.

Thus, even the design of the program itself promotes knowledge management activities, as both the process and the content are related to Knowledge Management. The participants learn about Knowledge Management while practicing it. The sequence of presentation, workshops, self-learning and transfer phases provides an optimal basis for learning.

Finally, the Knowledge Master certificate provides proof of the participants' knowledge, which should contribute towards the improvement of a particular management skill in their companies.

Knowledge Master's design and curriculum

Knowledge Master consists of several modules, each running from four to six weeks, with the full program running for six months (see Figure 45).

The *Basic Module* provides orientation and lays a foundation of thematic knowledge for the modules that follow. This module gives an initial exposure to and develops sensitivity for the complex interaction between the three knowledge management pillars, which are the focus of the authentic case studies.

The module begins with a workshop, where participants are given an opportunity to get to know one another and to form small learning teams. This initial module of the curriculum can be seen as a trial run for the virtual collaboration between the small learning teams (4-5 participants each), which will take place during the next six months.

To address the question of how knowledge should be mobilized, special attention is paid to the exchange and sharing of information and experiences. The second module, *Knowledge Communication*, concentrates on this, drawing on knowledge from the Psychology, Management, and Information Technology fields. Authentic case studies and accompanying literature require the participants to critically think about and solve problems. (For example: What features should a knowledge management system have if it claims to be both comprehensive and user-friendly?)

By having to form and support virtual teams, the participants also learn about group dynamics. They reflect on suitable incentives for the sharing of knowledge. They become familiar with obstacles in the communication of knowledge and ways to avoid or resolve these.

The last module, *Knowledge Management Tools*, completes the spectrum of important knowledge management topics, by introducing tools and methods of knowledge creation and utilization. The participants acquire a basic knowledge about available knowledge management tools, as well as experience, through an opportunity to test and use these tools. The idea is not to produce IT experts, but to develop IT literacy and the ability to better understand these important issues.

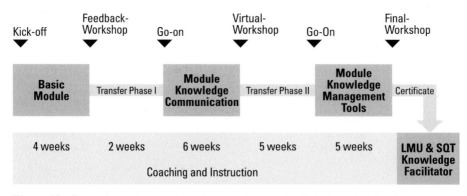

Figure 45   The design and curriculum of Knowledge Master (Principle) / Status: June 2001

There are three types of tools: technical (e.g. software programs that allow the visualization and graphical intertwining of knowledge areas); intellectual (e.g. implicit problem-solving heuristics that support the individual in his or her knowledge activities); and organizational (e.g. strategies that influence groups and/or the whole organization, in respect of knowledge-based changes).

All knowledge management tools are described briefly, giving the main function, possible applications, and critical success factors. The participants also have to document their experiences of the application of these tools. These documents resemble Best Practices or Lessons Learned techniques and thus preserve important knowledge about the successes or pitfalls of application. The documents are collected and stored on the Knowledge Web, and become inputs for and accessible by the whole community of learners.

One-day *workshops* at the start and end of each module round off the transfer of knowledge. In these workshops, participants take part in thematic discussions on the contents, with representatives from the three University institutes. The experts answer questions and the participants reflect on the learning and team building process of the completed modules.

Between the modules, there are two *Transfer Phases,* which foster the transfer of the learned material to on-the-job practice. Through structured questionnaires and instructions the participants are guided in their reflection on the learned contents and assisted in applying them in different contexts. They are also encouraged to form a learning community. The forming of a long-lasting learning Community by the participants is a very important element of the Knowledge Master curriculum.

The *Knowledge Web* is a web-based communication and collaboration platform that is offered to participants in the program, allowing them to create email-based discussion boards, chat groups, the creation of online-documents and the attachment of any other documents from disk. Participants are challenged to structure and organize this virtual space themselves. They have to find common rules and goals for a long lasting learning Community in a self-guided and democratic manner.

At the end of the further-education program, all successful participants receive a *Certificate* from SQT and the three Institutes of the University of Munich.

After each intake, thirty people leave the program qualified to initiate and manage Knowledge Management projects. They now possess a multi-facetted, broad knowledge base of the disciplines of Psychology, Management and Information Technology, having acquired methodical knowledge and competencies, and having undergone personal growth through their struggle and consequent success in the virtual learning environment.

Back to the conference floor

The lobby is once again beginning to fill with conference members heading for the coffee when Peter Müller returns to the Knowledge Master booth, to hear more about the actual course and plans for its future from Irmgard Hausmann.

"You know Mrs Hausmann, one of Knowledge Master's best features is the combination of theory and practice. We all liked solving the problems we were set, especially when we had to take the different perspectives on Knowledge Management into consideration. The fact that we were able to apply our newly acquired knowledge during the transfer phases and then discuss the experiences we had had during the next module was a great help. Being in a learning situation with university students was interesting, because they shed a different light on problems, which allowed us to see the blind spots our jobs had given us. Was this your intention?"

Irmgard is glad that he mentioned exactly the points that were the key elements in the didactic concept – the ones developed in the collaboration between SQT and the Institute of Educational Psychology.

Knowledge Master's didactic elements

Knowledge Master was developed by the Institute of Educational Psychology with the conscious application of the didactic principles of modern constructivistic learning-teaching investigations.

The result is a special combination of didactic elements (see Figure 46) that is followed in the Knowledge Master modules. Each didactic element is rooted in one or more didactic principles. A brief description of these principles and their transformation into the didactic elements of Knowledge Master follows:

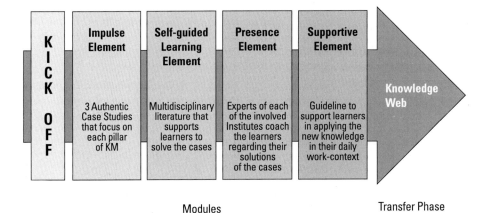

Figure 46 The didactic elements of Knowledge Master

Motivation, authenticity and applicability

Only when learners are interested in a topic, when they consider a given problem relevant and are consequently motivated to solve it, can they learn efficiently. How can their motivation be increased? Case studies appear to be the answer, as the learning content is embedded in an interesting context. As the case studies in the Knowledge Master are drawn from Siemens' wide experience in business contexts, they fulfill this requirement and increase the program participants' motivation.

Case studies also support the transfer of knowledge, adjusting the learning content to authentic contexts. In other words, knowledge that is learned in a classroom environment through listening passively tends to be inert. It cannot easily be applied in any other context. On the other hand, if the learning context is similar to real contexts, such as the daily work environment, it is much easier to transfer new knowledge from the learning situation to other, real situations.

In a nutshell: the more the learning context resembles the context in which the learned knowledge is to be applied, the easier the transfer of this knowledge will be.

How was this didactic principle realized in Knowledge Master? In order to motivate learners and to impart "vivid", applicable knowledge, the contents had to be presented in an authentic, realistic way.

Several authentic case studies, each describing problem situations in companies are presented in each of the modules as an Impulse Element. Each case study ends with a complex task that the participants have to solve co-operatively in small learning teams. The tasks allow plenty of opportunity for problem solving in that there is no "right" or "wrong" solution. While solving these problems, learners are nevertheless guided by concrete questions, hints, and tips.

Multiple contexts and perspectives

Another important factor that improves the application of new knowledge is the multiplicity of contexts and perspectives on the learning content. In other words, if new knowledge is viewed in different contexts and from different perspectives, it will be easier to transfer from the learning situation to real situations.

Each theory or knowledge element that is imparted to the participants in Knowledge Master is, therefore, viewed and discussed from the different knowledge management perspectives. This is achieved through interdisciplinary discussions at the various workshops, an integral part of each module in the Presence Element.

A distinctive feature of Knowledge Master, which contributes significantly to the multiplicity of perspectives, is the formation of small groups, consisting of both employees and students. While the students become acquainted with the practical problems and problem-solving strategies of the employees, the latter benefit from the theoretical approach which the students bring to the joint tasks. The employees also come in contact with new scientific theories that enrich their problem-solving strategies.

Social learning arrangements

One of the golden rules when designing a learning environment is "learning by teaching". When learners are asked to explicate their newly acquired knowledge to other learners, they have a much higher success rate than when learning silently. Learning in a social setting also makes provision for the many perspectives that exist within each group. Each member comes to the group with different previous knowledge and experiences and therefore, with a different perspective on the learning contents. In discussions, these diverse views and exchanges of previous knowledge and experiences result in each team member profiting from the others. The results are elaborated in team solutions to complex problems that surpass the solutions of individuals.

Knowledge Master's design is strongly focused on social learning arrangements. The small learning teams, formed in the very first Kick-off Meeting and together until the end, play a vital role. While most of the teamwork is done virtually on the Knowledge Web, team members also meet at the workshops and in the interim, if they so wish. The Knowledge Web offers them several possibilities of communicating and collaborating by computer. (A more detailed description of this appears further down.)

Self-guided learning

In recent years, those involved in designing instructional programs have increasingly advocated self-guided learning. The typical learning-teaching environment, found in the school situation, is no longer considered ideal for learning complex, multi-facetted knowledge. A higher degree of freedom can be achieved in a learning environment if learners themselves are allowed to decide what they learn, with which materials, in what timeframe and in what depth.

When learners are new to a specific knowledge area and cannot draw on previous relevant knowledge or experiences, too much freedom is counter-productive, as the novices are overwhelmed. However, more experienced learners are more motivated, and thus more successful, through self-guided learning.

The didactical design of Knowledge Master has made provision for a considerable degree of self-guidance in the *Self-guided Learning Element*, but also offers supportive guidelines for those who feel overwhelmed by the task of organizing their own learning process.

The *Self-guided Learning Element* offers background literature to help solve the problems set in the case studies. The literature consists of empirical studies, newsletters, or business guides, which the participants can use if they are not able to find their solutions in other sources.

The learning teams can decide themselves how they want to solve the authentic case study problems, for example, whether they want to divide up the tasks or make the solution a group effort. The form of the solution is not predetermined and participants may design a PowerPoint or video presentation, or prepare a written text.

Instructional support

Closely related to the degree of self-guidance in learning environments is the role played by the facilitator. In a typical classroom setting, the teachers present pre-structured knowledge and largely guide the learners. The teacher determines what is to be taught next. In authentic, problem-based learning environments like Knowledge Master, the teacher's role changes radically: The teacher becomes a facilitator who accompanies the learners on their way through the complex learning environment. The facilitator's role is to support learners to find their own answers. Learners have to actively discover new knowledge and link it to their previous knowledge. The facilitator has to offer support in this process, when necessary, or when the complexity of the new knowledge makes it necessary.

In the Knowledge Master program, moderators from both the University and business, tutor participants. A first-level support provides the opportunity to contact moderators by telephone or email. They then observe the activities of the teams and give hints, provide further literature or other forms of support, if needed.

Besides the moderators' presence, the Supportive Element in Knowledge Master also assists learners. This is a frequently updated document that lists relevant information; such as further training possibilities, contact persons at SQT, or interesting resources, or newsgroups on the Internet. It also offers tips on how to apply the learned knowledge in the workplace. It offers a step-by-step guide and explains the evaluation of each learner's learning progress and provides instructional support for self-guided learning and working. Finally, it offers strategies and methods that may prevent information overload and stress in the workplace.

At the conference again

At her booth, Irmgard Hausmann is preparing to show Peter Müller the new Knowledge Web.

"Do you know we actually did take the suggestions that your group made after the pilot program to heart? We designed a completely new user interface. You won't recognize it. Let me show you".

She is proud to explain the redesign that was done by Prof. Mandl and his colleagues, after the evaluation of the feedback from the pilot participants. "We decided that we wanted a platform that is as easy to use as possible".

"That must have been an expensive exercise", Peter comments.

"Not really, but user satisfaction is infinitely more valuable to us than the cost of redesigning. The users need no longer worry about operating the program. They can simply concentrate on the contents of the program and its communication".

The learning platform "Knowledge Web"

The virtuality of the Knowledge Master curriculum is only interrupted by a few face-to-face workshops. A technical platform is thus needed to provide the participants with all the relevant information and allow them to communicate and collaborate by computer.

The *Knowledge Web* meets these requirements. It accompanies the participants through the curriculum, during which it serves as a communication and collaboration platform, a portal for all relevant information and a shared workspace. The Knowledge Web is a restricted area at the SQT Internet site, to which only participants and moderators have access. It is based on comprehensive web technologies, allowing users to access it from anywhere with their Web browser without the additional installation of special client software. It contains all the information needed by the participants of Knowledge Master and is frequently updated. It is supplemented by another very important aspect of the Knowledge Web – *the Web Board* – that serves as a virtual team room.

A brief description of the features of the Knowledge Web follows.

The entry page

When entering the Knowledge Web, an *entry page* appears, which serves as a single point of information. It announces administrational changes, provides news and updates on the team tasks, and contains all new documents presented or generated at the workshops. Presentations by experts, or photos of meetings are also found here. It is a portal that connects to all other sections of the Knowledge Web (see Figure 47).

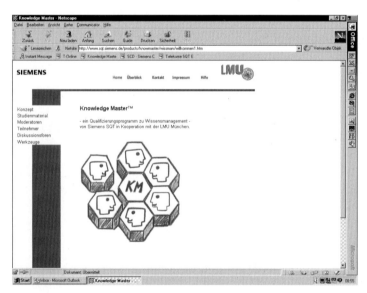

Figure 47 The entry page of the Knowledge Web

The navigation

In the *Concept section*, the background of Knowledge Master is given. This section also displays the current phase of the curriculum, describes the relevant learning tasks and their deadlines.

The *Study Material* section may be the most important one. It contains all the learning material: the authentic cases, the tasks and background information necessary to solve them. In this section, comprehensive abstracts of all the relevant literature are provided for downloading. The teams' solutions to the problems raised in the case studies are also displayed here. All case studies, all tasks for the cases and the corresponding solutions are all accessible from the same screen.

The *Moderators section* introduces the Knowledge Master moderators: Each name is linked to an email address, and a telephone number, postal address, a photo, a short description of the moderator's competencies, role, function and some personal details are provided.

The *Participants section* displays a photo of each participant, his or her contact information, and introduces the person briefly. This information is obtained from the initial meeting when all participants are required to write a few sentences about themselves. Direct email contact with all the participants is also possible from this screen.

The *Discussion Forum* takes the participant directly to the Web Board, which will be described more fully later.

The final section, *Tools*, contains a download area for useful software like WinZip, Netscape Navigator, Adobe Acrobat, etc. The intention behind this section is to allow participants to access Knowledge Web from elsewhere.

The header navigation

In the Knowledge Web, the most important functions are always available at the top of every screen. Clicking on the Contact button opens an email form that sends messages directly to the moderators of SQT. The Help button leads to a detailed description of all the Knowledge Web features. The other functions on the header navigation are:

- a Site Index that allows the user to navigate directly to the information needed,
- a Home button that returns the user to the entry page, and
- the Site Index that lists the names of those responsible for the contents of the various sections.

The Web Board

The *Web Board* serves as a meeting point for virtual communication and collaboration (see Figure 48), offering each team a private forum where its members can share their opinions and work on solving cases.

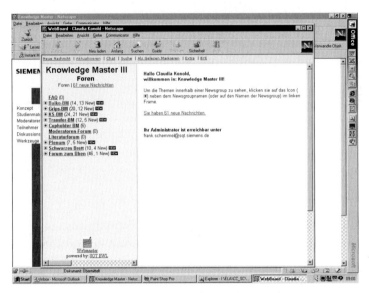

Figure 48 Screenshot of the Web Board, KM III

Besides these private forums, there is a Plenum and a Notice Board, which are both open to all participants. The Plenum is a discussion forum where general questions about the case studies and other interesting topics are discussed with all participants and moderators. The aim is to connect the small teams to a whole community that virtually ponders interesting topics in the knowledge management field. The Notice Board organizes the virtual phases. It serves as reminder of the dates for forthcoming meetings or deadlines for team solutions. Here participants may also discuss topics that are not relevant to the solution of case studies.

Each case study also has its own thread in the Web Board to which the teams post their solutions, so the required information is easily and readily accessible.

The Web Board stores all messages, thus providing the participants with a full documentation of all communication. This function may be used in the later phases of the curriculum, to review solutions of problems set in earlier modules.

The Web Board allows for the attachment of any type of document to a message. This means team members are able to send their contributions to solutions of problems as a document to the team forum. There it may read by the other team members and re-edited. Eventually they will arrive at a common team solution.

Participants can also chat with one another on the Web board, although experience has proved that this function is used only at the start of the curriculum. This may be due to participants appreciating an informal chat at the beginning of the team-forming process, but later, when complex communication on studies is required, a chat line no longer satisfies them. Email and conference calls then become the preferred communication media.

The Knowledge Web has increasingly become Knowledge Master's communication hub. Once people know each other, they communicate, learn and work together successfully in a virtual learning environment.

Looking back at the end of the conference day

During the lunch break, many prospective Knowledge Master participants visit Irmgard Hausmann's booth. Many former and current participants wishing to speak to her also come past. She invites these "graduates" to join her for a chat at the end of the conference day at the restaurant.

After the last presentation, five people join Irmgard at her table in the restaurant and introduce themselves. They may have not worked together on the same virtual team, but they share the same context and understand each other immediately. They spontaneously start developing ideas on improving their knowledge management activities, ways to help one another, and on collaborative initiatives.

But Irmgard is also interested in finding out what they consider as Knowledge Master's most important contribution to their lives. The answers differ quite widely.

"The mix of computer-enabled learning and face-to-face workshops is a powerful combination that I haven't found in any other further-education programs. Face-to-face training would not have been possible for me, as I could not have left my job for six months".

"I found the program's practice orientation most important. Knowledge Management is normally taught theoretically and I had previously attended many talks on the subject, but always missed the link as to how all that theoretical knowledge should be transferred to the job. I found the authentic case studies used in Knowledge Master a great help".

One of her guests comments: "The self-guided learning method was a huge challenge for us as individuals and teams. It was not easy to organize collaborative work, especially in the first weeks when the virtual communication in the team was not yet working well. The Knowledge Web was very good for the presentation of our results, but the workgroup performance was not always on a par. The time allocated for the solutions of tasks was also rather short. After having studied the literature, we only had a few days left to develop a common solution within the team. This was a great challenge, but somehow we managed it".

Another adds: "Yes, with the support of our moderators and teletutors, we finally all became good teams and learned a lot about one another, even if we did not meet very often. This type of further education was perfect for me, since I could immediately transfer lessons I had learned to my job – often the very next day! That was quite unique. I have recommended the Knowledge Master course to some of my Knowledge Management colleagues".

Irmgard is very pleased to get this feedback from the former participants. "Well, your impressions have been confirmed by the evaluation that was done by the Institute of Educational Psychology to optimize Knowledge Master. Do you remember the questionnaire you filled in after the basic module? This was one of several steps in the evaluation process. The results of that questionnaire showed that the basic module is a very suitable introduction to Knowledge Management. Nearly all the participants said that they enjoyed the work very much, especially the case-oriented learning. They felt that the cases demonstrated typical problems in the company's daily reality and that, in combination with the tasks, literature and face-to-face meetings, they offered sound knowledge about the goals and main components of Knowledge Management".

Everybody is very curious to hear more about the scientific evaluation and Irmgard continues: "Not all of it was good news. The results also showed that the participants should have received more support and instruction, and that the learning material should be reduced. The participants would have liked things like an example for the solution of the tasks, which would have given them hints about the form, depth and length of the solutions. Although they found the tasks for the cases and the supporting literature lucid and informative, they also found them too extensive. In later trials, these recommendations were implemented and the concept was altered to include more instruction and better prepared literature".

Later that evening, Irmgard Hausmann drives back to Munich. The conference was a great success. Not only had there been considerable interest in Knowledge Master, but some participants had already reserved a place in the next run. She would also be lunching with the Vice President of Human Resources of a large financial services company the following week. He was looking for a program that would allow him to better support the development of knowledge networks in the sales department of his organization. He thought that Knowledge Master might just be the answer.

Irmgard is tired after the hectic day, but pleased about the positive responses from the former participants. However, her to-do list has grown, as she now has even more ideas for further improvements to Knowledge Master.

She takes the exit road that will lead her home and simultaneously makes a decision: Next year, she will invite everyone who has already received a Knowledge Master certificate to the first Knowledge Master conference. Yes, the network for the exchange of information and experience has been established and will continue to expand.

Looking back after a year

About one year later, at another Knowledge Management conference, several previous Knowledge Master participants met during a break in the lobby. As Irmgard Hausmann and some people of the University of Munich joined the group an interesting discussion began: "What about Knowledge Master, is it continuing?", several asked. "Yes, sure"!

Irmgard answered: "We have already had more than 150 participants in 6 courses, and we are just preparing no. 7". "How was the Knowledge Master conference you planned in springtime last year? Unfortunately I could not come because I had a project in Malaysia". "Oh, it was great" some others replied, "there were about 80 of us – we could visit various show cases on subjects like "Story Telling" or "KM-Metrics". "But the exchanging of experiences was the best part of the conference", another said.

"By the way", Irmgard mentioned "Knowledge Master is going international"! "After the implementation of various practice oriented improvements for the next run, we are planning an international version in English. Yes, Knowledge Management isn't going to stop at the political or cultural boarders. But to overcome them requires a good concept, a lot of global experience and a network of best partners. We are all looking forward to this and great new challenge". As the conference started again, the group decided to continue the interesting discussion in the evening.

Key propositions

Based on the personal experience of the Knowledge Master creators and a systematic evaluation of the didactic concept by the Institute of Educational Psychology, the following statements reflect the most important lessons learned in producing a successful corporate further-education program on Knowledge Management.

1. Knowledge Management is a multi-facetted knowledge area. Anyone who wants to manage knowledge must know about the psychological, economic and technical implications of each activity involved. A knowledge management training-program should therefore always try to provide a very broad perspective on knowledge management activities.

2. Knowledge Management needs an interdisciplinary design, as it relies on very different, but important perspectives, namely the human, business, and technological perspectives. Collaboration with the Institutes of Educational Psychology, Management, and Information Technology at the University of Munich allowed us to bring in leading expertise in these fields.

3. Learning about Knowledge Management should not end with the learning of theoretical knowledge. Being a knowledge manager means having "lived" experiences in project management, in face-to-face and virtual teamwork, in cultural and personal barriers to sharing of knowledge and so on. A learning environment that claims to teach Knowledge Management should therefore try to simulate the environment in which Knowledge Management is applied. The participants in Knowledge Master, therefore, not only read about a typical knowledge management problems like the communication barriers in virtual teamwork, but they directly experience these problems as a result of the authentic, problem-oriented learning environment!

4. A further-education program can only be said to be successful when the participants can apply the learned content in real work settings afterwards. The main goal of a further-education program should therefore be the transfer of the learned material to reality. This can be realized through the concept of case-oriented learning in an authentic learning environment, as is done in Knowledge Master.

5. A further-education program that is mainly virtual (like Knowledge Master) faces very particular challenges. The virtual learning environment requires a sensitive approach, detailed attention to all the steps in its development, as well as during the whole curriculum. Its conception and design have to match the didactic design closely. The expectations of the participants must be taken into consideration. Each time the six-month program is followed, the technical learning platform needs to be constantly updated and special attention must be paid to the facilitation of the communication on the Web board.

6. The virtuality of Knowledge Master requires special competencies of both trainers and facilitators alike. They must be face-to-face moderators and teletutors, and consider the participants' expectation and problems. They should also provide detailed feedback on the participants' solutions to problems. This requires a thorough knowledge of the content, as well as psychological and communicative competencies. In a virtual learning environment, tact is vitally important.

7. A good further-education program should combine theory and practice in order to convey new theoretical knowledge, whilst simultaneously ensuring that the selection of this theoretical knowledge is relevant to the practice of Knowledge Management. The more the learning environment resembles real circumstances of the workplace, the more authentic it is, the easier the transfer of the learned contents will be. Another method of combining theory and practice is to mix students with employees – both parties benefit from the other's presence, as Knowledge Master has proved.

8. A further-education program may be brilliant, but still not be accepted. During its design, it is therefore essential to take into consideration the expectations, problems and desires of the potential participants, even if it means that the program differs from what was intended. Knowledge Master' flexibility was a feature that made it very acceptable: participants may work on it for just one hour a day, if necessary, while continuing to work full-time. Changes should be made in response to the findings of evaluation studies – in Knowledge Master's case the learning material was reduced. A good further-education program must compromise between the didactics designer's ambition to teach very profound content and the practical requirements and constraints of the working environment.

9. The best way to determine the nature of this compromise is to make a formative, scientific evaluation. An ongoing evaluation study, running parallel with the program, is essential for pinpointing problems and difficulties right from the start, so that adaptations can be made as soon as possible. Formative evaluation helps to optimize the program and leads to transferable results – which can be used in other further education programs. Finally, the implementation of a program concept will always differ from what was originally expected and planned. Start and adjust it as you go along.

10. The planning of a further-education program for a broad subject like Knowledge Management requires many resources and competencies. Knowledge Master's success is attributable to its many different elements: the self-guided learning, the authentic cases, the interdisciplinary nature that offers multiple perspectives, the social learning environment, the virtuality, and the instructional support. One of the core elements for planning such a program is a sensitive, didactic concept that integrates and balances all these elements.

Discussion Questions

1. Sketch the requirements of a corporate learning program, in general, and then, those that are required by a program teaching Knowledge Management, in particular.

2. Discuss alternative designs for a further education program on Knowledge Management.

3. Discuss the strengths and the weaknesses of the design of Knowledge Master's modules.

4. Discuss how well didactic principles have been transferred to Knowledge Master's modules.

5. Where do you see potential for improvement?

6. Make suggestions for short-term and long-term steps to develop Knowledge Master further.

The Siemens Management Learning Program

Christina Bader-Kowalski & Antonie Jakubetzki

Abstract

This case study comprises a general survey of the background, approach and structure of the Siemens Management Learning programs. Founded in 1997 these programs marked a radical new beginning in management education. Today they have become crucial for knowledge management within the company and represent the most important force for worldwide strategic management development within Siemens. The case study, moreover, provides several examples illustrating the procedure and results of the Business Impact Projects that form – beside workshops and e-learning – a major part of the programs. The build-up and purpose of the projects are first explained, where after concrete examples as well as business results and key learnings are provided. Finally several key propositions enhance the transfer possibilities to other business contexts.

Background

It is not easy remaining competitive at a time when business processes are changing rapidly and agility in responding to customers is crucial. Siemens, however, has nothing to fear. After realizing that its business was increasingly based on knowledge, it identified its employees' knowledge, talents and skills as the decisive factor for success. This success factor will also ensure the company's competitive edge in future. To guarantee this future success Siemens will make it a matter of course for everybody within the company to exchange their experience and practical knowledge swiftly and simply, so that everyone in the company can have access thereto and use it in their everyday work.

Since knowledge is mainly built up and exchanged within a personal relationship, the basis for the above-mentioned knowledge management is, above all, trust. This is where our managers play a decisive role. They drive the exchange of knowledge. In daily dealings with their staff and with their involvement in projects, they create the required trust that continues to push this Best-Practice thinking further. However, their ability to fulfil this demanding role cannot be taken for granted – it requires a highly educated staff. If managers are to play a key role in knowledge management, they need all the active support, training and education their company can provide.

Figure 49 Overview of Siemens Management Learning Programs

In 1997 Siemens responded to these requirements by establishing the Siemens Management Learning program. During the past four years this program has become an essential component of the knowledge-based organization. Worldwide the Siemens Management Learning programs represent the most important force for developing Siemens managers' skills.

Siemens Management Learning's concept comprises 5 programs, ranging from S5, for potential managers, to S1, for top management, with each building on the other in terms of content (see Figure 49).

The main advantage of Siemens Management Learning is that throughout the world all the management learning programs have the same basic structure. S5 and S4 are run locally in the relevant language, whereas S3 to S1 programs, which are run regionally (S3) and globally (S1 and S2), are only presented in English. Similarly all programs worldwide are based on the same learning methodology and program structure.

More than classroom instruction

In the past, management seminars were often simply lectures with very little impact on a firm's day-to-day problems. Siemens Management Learning programs have contributed specifically and measurably to business success.

With the introduction of the programs, Siemens risked a radical new beginning by:

- replacing the "seminar" method of training with a more efficient management development system;
- offering programs that are a mix of cross-cultural exchange of know-how, networking and active project work that prepare managers thoroughly and practically for their role as leaders, and

Figure 50 General structure of the Siemens Management Programs

- extending the duration of the programs to a whole year. In this way lasting changes in the attitude and behavior of our managers can be guaranteed, and management expertise can be thoroughly established. It would be foolish indeed to believe that lasting effects can be achieved within a week.

Globally the Siemens Management Learning programs are designed around three basic parts: Workshops, E-Learning and Business Impact Projects (Figure 50). These parts are individually discussed below.

a) Workshops

The workshops of the Management Learning Programs are focused on increasing business knowledge and leadership competencies as well as Siemens' current initiatives. To enrich internal knowledge with the views of outside experts, the workshop participants also collaborate with professors from various business schools. Duke University, Babson College, and St. Gallen are just a few of the outside partners that help stimulate new thinking and new approaches.

b) E-Learning

E-learning is also a major part of the Management Learning Programs. Starting with the preparation phase, participants are encouraged to use self-study material, since e-learning takes place in the period between the workshops (see Figure 50). E-learning is a very effective way of learning since it enables the participants to cover the course material when and where they wish to do so.

A source of learning material that all participants should access, is the intranet-based Learning Landscape, which is an e-learning platform for Siemens-specific knowledge. It provides the participants with interactive learning modules as well as covering a wide range of general information. The Learning Landscape therefore represents an ideal platform not only for the management learning participants, but also for all Siemens employees wishing to enhance their knowledge.

Besides the self-study, virtual teamwork is also required during the various periods between programs. To this end, state-of-the-art e-collaboration tools are available that provide an on online environment for e-meeting events with voice communication, application sharing, online breakout session with electronic flipcharts etc.

c) Business Impact Projects (BIPs)

A fundamental goal of Siemens Management Learning is to combine learning and business. Each program is based on the belief that managers can benefit most from new learnings when they put those learnings to work to produce new results and greater accomplishments. To this end a participant who would like to successfully complete a program, has to work on a BIP (Business Impact Project), which is a short, result-oriented project that will have an immediate, measurably beneficial effect on some aspect of our business.

The participants contribute the ideas for these projects themselves, they are, however, asked not to limit the range to the business connected with their present position. The best of these ideas are finally worked out in cross-functional teams of 4 to 6 members each. Finding a suitable topic for a BIP is a primary challenge, since a BIP has to fulfil 2 main criteria (see Figure 51 for an explanation of the dimensions of a BIP).

Business Impact Projects should

- firstly be linked to a *business opportunity* – a potential long-term gain for Siemens – and
- secondly, illustrate the proposed long-term gain in a specific situation by achieving measurable results in a short timeframe, which is summarized by the term *breakthrough goal*.

In practice the first step of a BIP team is finding a business opportunity. Within the scope of a potential long-term gain, the BIP team focuses on a specific situation offering a breakthrough goal. This goal has to be attained in the limited period available for the team's BIP work. The measurable results must furthermore explore the potential for attaining large gains for Siemens in the overall business opportunity.

To meet these challenges, each team is supported by a coach. The coach or client is normally a senior manager within Siemens who has a genuine interest in completing this project that could make a concrete and measurable contribution to his business unit's

Figure 51
Dimensions of a Business
Impact Project (BIP)

231

success and therefore to the corporation's business success. This person not only supports the team, but is also responsible for the implementation of the project.

In its entirety, the BIP work is a great challenge for the participants who are faced with many problems, the greatest of all being time pressure. Each team has no more than 4 to 6 months to spend on implementing an actual solution. In addition, the BIP work demands discipline and diligence. The BIP work phase not only comprises the solution-finding phase, it is preceded by problems such as finding a topic, a coach, and drawing up a plan. On the whole, the BIP keeps the participants busy throughout the timeframe. This is an extremely arduous undertaking if one considers that it has to be achieved parallel with a participant's regular day-to-day activities.

The knowledge management platform

To organize and document all running and completed Business Impact Projects, Siemens Management Learning developed an intranet database called the BIP Marketplace. The BIP Marketplace is the forum that brings participants of the Management Learning Programs, business ideas, projects and coaches together. Through the BIP Marketplace, the management learning participants have access to all relevant data concerning both their program and their colleagues.

The BIP Marketplace contains 2 points of entry: a public access and a participant access.

The public access

Via the public frontend of the BIP Marketplace, every Siemens employee worldwide can obtain information about all currently running Business Impact Projects in the Siemens Management Learning Programs. Knowledge is shared through the BIP Result Database, an intranet–based knowledge management platform where the documentation of all completed BIPs is available to the company as a whole. After a BIP has been completed, it is published in the BIP Result Database. A template allows the participants to access previous BIPs' project results, outcomes and lessons learned. By making these public in the database, project members share the results of joint projects and their personal know-how with others.

By sharing the experience company-wide, employees with similar problems can profit by making use of this know-how and solutions.

The participant access

In addition to this public information, management learning participants have their personal "knowledge pages". If the public pages only contain general contact addresses, this second access provides the participants with detailed contact information as well as an individual competency and skill profile of each participant. With these knowledge pages it is easy to find the accumulated knowledge (e.g. languages, IT know-how etc.) of all management learning participants. All participants can update their personal profile and add their individual competencies. Knowledge pages make it easy to identify and find a competent partner within Siemens.

The straightforward and user-friendly structure of this concept can be seen from the many successful BIPs documented in the BIP Result Database. This knowledge pool of Siemens Management Learning forms the basis for a learning company and makes a valuable contribution to company-wide best practice sharing.

Examples of successful Business Impact Products

The Safe Ship

Siemens Building Technologies (SBT), Fire Safety Division is currently a market leader in the safety and security business. In 1999 a BIP team decided that their breakthrough goal would be to convince the Fire Safety Division that the marine business would be a profitable new segment for them.

As demanded in the BIP work process, the project team had through thorough contemplation, first identified a general business opportunity for Siemens. According to this team, Siemens could create added value by expanding its market share in targeted market segments. To reach this long-term goal, Siemens had to specifically take advantage of the profitable after-sales business system, as well as the application of existing knowledge within the corporation in order to gain a competitive edge.

But how could these large gains be achieved in a limited 4-6 months period? Within the BIP work it is fortunately "only" necessary to set a good example in a concrete situation. BIP teams, for instance, achieve measurable results within one division, or in cooperation with one specific business partner. However, by attaining this goal they seek to illustrate the potential of an overall business opportunity for Siemens.

This BIP chose the Fire Safety Division to be the shining example. This division had always concentrated on developing fire detection systems in buildings which, the BIP team felt, lent itself naturally to expansion to the marine business. The BIP team was sure that the Fire Safety Division could become the second largest provider of marine fire detection systems within 4 years. They subsequently created a professional and well-structured business plan for the division's entry into the marine sector. This finally convinced the Fire Safety Division to implement the BIP strategy.

Results of the initiative

In 1999 the BIP team launched its first pilot project with TT-Line, a German/ Swedish shipping company, as their partner. This had come about after the BIP team had used existing contacts to the shipbuilding industry within Siemens – which goes to prove how much insider experience helps one to obtain an edge! The Fire Safety Division then delivered the Cerberus® fire detection solution for two newly built ships of the TT-Line that were to undertake their maiden voyage in the summer of 2001. Apart from these two ships, owners of some 20 other ships – mainly carriers and tankers, but also special installations such as drifting oil rigs – have requested that the existing fire detection systems be replaced by the Cerberus® solution from Siemens. There are many more exciting business opportunities awaiting this new Siemens product.

On the long term this BIP was especially useful, because the team created a realistic reference: They proved that a new Siemens entity can generate profitable business by utilizing existing Siemens sales channels. It is now up to other divisions to follow. And follow they shall: In the security business new concepts are already on the drawing board.

According to the team it was all great fun, but hard work and an enormous challenge. The latter comes as no surprise if you consider that there was only one expert on fire detection systems in the team, that the team members were from different locations and that they were under tough time pressure. However, by ensuring that professional experience triumphed over personal interest, that the focus was strictly on achievable goals (they focussed only on fire detection), through their rigid project management and, finally, by ensuring a clear structure and methodology, the project ended very successfully. In addition this was a team who communicated efficiently and openly and who was fully dedicated to "The Safe Ship" BIP.

Creating additional UMTS revenues

UMTS (Universal Mobile Telecommunication Systems) is the next generation mobile phone network. The challenge for one S3 BIP team, who came together in 2000 and finished their project in July 2001, was to identify ways to operate a UMTS network for one or more licensee holders – sharing the network – thereby reducing costs dramatically. The investment needed from operators to build a new network is, however, immense (approximately $2 billion for a midsize European country), but if Siemens could achieve the goal of this particular BIP, it would make the corporation more attractive as a UMTS provider by offering reduced entry costs, particularly for new operators.

With Siemens actually operating parts of the network, it could receive payments from the operators who run these operating services. For Siemens this would mean new revenue streams. The additional challenge facing this BIP team was that operating networks was specifically excluded from Siemens' business strategy, therefore any solution which included such an idea would be contrary to corporate strategy. This meant that as a ground-breaking idea there was no backup in the form of market research, and that the regulatory environment and license conditions had not foreseen any attempts to reduce UMTS investment costs, for example, by sharing infrastructure.

The team created a team structure and a project plan. Major project goals were, for example, to:

• establish a coach and sponsor support,

- define possible concepts/model building and offers to make to operators (e.g., network infrastructure sharing),
- locate suitable potential countries and customers to develop the concepts (in joint workshops), and
- ensure good internal support within Siemens.

The overall goal was to conduct a joint-workshop with a customer to develop the concept into an implementation plan.

Results of the initiative

Impressive results were achieved. A properly implemented BIP model is financially more attractive than a traditional vendor finance approach and provides a win-win situation. Understanding this market option before the competitors did, was also important. This BIP increased the IRR (international rate of return) for future multimillion dollar projects and generated additional revenues for UMTS projects.

Within the scope of the BIP, the team learned about managing virtual teams and using new technologies. Running a virtual team who is geographically separated and whose members have full-time jobs, requires considerable commitment and discipline. The team created and adhered to a team contract that constituted an agreement on how to best implement this project. The commitment and support of the coaches and sponsor were very valuable and extremely important for the success.

The BIP team identified a potential substantial revenue improvement for Siemens and the BIP idea was accepted internally as worthy of following up. Sales leads are now being followed through.

The finger print system – developing a market for biometric sensors

The BIP team of one S3 Seminar, starting in the spring of 2000 and finishing about one year later, decided that the development of a whole new market would be their business opportunity for Siemens. The specific target was developing a market for biometric sensors with sufficient business volume to generate a positive EVA (Economic Value Added).

At this point of the project, the members of the team had to answer questions such as: "Why are biometric sensors such as the fingerprint being used?", "Is it possible to develop a market for biometric sensors?", "Are these sensors truly to the customer's advantage, easy to use, and will no losses etc. occur?" and "Are the customer advantages enough for biometric fingerprints to prevail in the marketplace where they face competition from the non-biometrics?"

Following their goal, the S3 team developed a business plan that presented the broad range of security applications that were possible products for product development companies, as well as manufacturers of industrial equipment and automation and system suppliers.

Soon it turned out that the project – due to its enormous complexity – required the cooperation of 3 Siemens groups: Siemens ICM (Information and Communication Mobile), Siemens AT (Siemens Automotive Technology) and Infineon. In this initial situation, however, there was no common solution approach between these Siemens divisions. In the interest of the business opportunity, the BIP team achieved a combination and optimization in the cooperation between these Siemens groups by focussing on this one goal. The work and mediation of the team lead to a written agreement between the concerned groups. This cooperation proved itself so profitable that it will be continued even after the BIP has been completed.

After improving the capacity to act, the team contacted two customers from two different market segments who intended using fingerprint technology or systems in their products.

Results of the initiative

A signed letter of intent to use fingerprint technology or systems was the main result of this BIP: from Audi there is interest in fingerprint technology in a car, while Cherry is interested in fingerprint technology on a PC keyboard.

On the whole, the BIP showed that there is indeed a market for biometric sensors. A further outlook is that Siemens AT is very interested in the fact that the non-automotive fingerprint business is expanding rapidly, since this would allow it to participate in the economics of scale for later introduction. This may lead to ongoing monitoring of the situation to guarantee the future AT profitability of fingerprint applications.

After working on the "Development of a market for biometric sensors" BIP, the flexible and international team of 6 members was able to sum up the learned lessons as follows:

- A good sponsorship is a huge enabler. The involvement of the sponsor increases the chances of a positive project result.
- The importance of a clear to-do-list for all team members at the end of a meeting cannot be over-emphasized.
- Additional help from outside (for example, that of trainees) can maintain progress and help a team to keep to the time schedule.
- Big items should be divided into small projects with measurable results.
- It also was a great challenge to unite the efforts of several business units that had never previously cooperated.

The mobile phone cost-saving initiative

Siemens AG makes 12 million mobile phones a year, but its own use of mobile-phone service was in need of review. For example, managers in different Siemens units in Great Britain were buying phone services for thousands of employees from a number of suppliers, instead of working together as a team to negotiate a cut-price contract from one source. The internal administration of the providers' services was also inefficient and they were not making the best use of the technology (the user manual of one of the mobile phones hadn't been updated since 1991). It was at this point that an international, creative and flexible team of 6 Siemens participants saw an opportunity to save the company significant money. This challenge was selected as a BIP for the S3 Management Learning Program.

The team was responsible for developing a strategy to advance the initiative and then break the project down into manageable modules so that short-term goals could be quickly achieved. They identified three areas of impact (mainly focusing on the UK) that would lead to significant cost reduction: Smart Purchasing, Smart Usage and Smart Administration.

Results of the initiative

Significant cost reduction and EVA results were realized as a direct result of the team's work. In September 1999, twelve months after the formation of this BIP team, lower call charges of 1.5m Euros were banked. During the two financial years ending in September 2001, further procurement, usage and infrastructure savings of 8.3m Euros will be realized, and even greater savings are expected in future. Financial benefits were also gained from sensible usage of the phones; for example, making calls using the newly introduced mobile network, awareness of different tariff charges, personal payment for private calls, and direct feedback via a resilient billing process to ensure that users actually see how much they are spending.

Other benefits will also flow back to Siemens as a direct result of the team's hard work, such as having the flexibility to make private calls. Management information will be provided by the service provider which will allow Siemens to review call patterns and identify further opportunities to reduce costs. Ultimately Siemens will be freed from the administrative burden of managing the billing process and add real value by identifying external business opportunities utilizing the new mobile phone technology.

Conclusion – what have we actually achieved?

- We have reduced the complexity of the learning systems and introduced a clear program structure tailored to future needs, such as knowledge management and virtual team processes.
- We have demonstrated and measured the business contribution of learning by working on concrete business tasks in a clearly structured, project-oriented learning environment.
- We have promoted Knowledge Management across functions, groups, and country borders and have we established networks among managerial-level staff.
- We have achieved progressive internationalization through uniform world-wide programs that can be adapted to local and regional requirements.
- Yes, we are important supporters of Knowledge Management – a driving force guaranteeing competitiveness and the success of Siemens!

In addition, a number of new cultural elements of management development and learning were consciously introduced into the company through Siemens Management Learning. These were the orientation of learning processes toward performance and results, the use of virtual learning environments, the worldwide exchange of knowledge using the latest technologies and the integration of learning and work, just to mention the main aspects.

Furthermore? With each BIP of a Siemens Management Program added value can be created for Siemens. Cost are reduced, delivery time decreased, customer satisfaction improved … All this is done by participants who had the chance to develop themselves and to enhance their knowledge on a wide range of areas.

To date approximately 5000 managers have already taken part in our programs and the demand for our programs is still growing. There are 1200 running programs and the implemented BIPs are evidence of our success.

Key Propositions

1. **A successful management learning program consists of different parts.** Workshops provide the participants with state-of-the-art knowledge and working tools. Besides classroom learning, e-learning is the perfect medium to enhance the acquired knowledge and to equalize all the participants' knowledge standard. Thus the time and location of further education can be individually determined by every participant. Finally the theoretical input has to be transferred into practice by working on real time projects. The duration of the program should be a minimum of a year. Within this period a reasonable timeframe has to guarantee the linking of theory and practice in order to provide a useful learning situation and appropriate transfer possibilities. An informal atmosphere within workshops and the whole program is a precondition for socialization between the participants.

2. **The goals of a management learning program should support the development of division-spanning networks.** Real time project goals and focal points outside the day-to-day work of the participants, enhance the growth of division-spanning networks. The unbiased approach to problems outside their daily work, gives each member the opportunity to act as a consultant. In addition, the challenge of improving knowledge and skills not only benefits the participants themselves, but also the human resources value of the company.

3. **The facilitator of the program should support the formation of international teams.** In addition to the general lessons to be learned in e-based teamwork, multicultural teams provide benefits that are not easily obtained in other situations. Beside the potential to improve their language skills, the members of an international team are encouraged – through practical experiences – to increase their intercultural competence. Through the combination of different vantage points – as a result of differing cultural backgrounds and educational systems – multicultural workgroups provide circumstances that encourage innovations.

4. **The facilitator of the program should ensure the encouragement of a coach/client.** The facilitator supports a team by searching for an appropriate coach for a project. The business experience, time at his or her disposal, and the motivation and responsibility of the coach have to be taken into account in order for the team as well as well as the coach to benefit optimally. The major task of the coach is to ensure the success of the project and he or she usually occupies a strategic position in the business unit at which the project goal is aimed. The coach has the business influence and contacts that are needed to achieve the project targets within a limited period. However, he or she is basically the client of the team, although to ensure success he or she, with the support of the facilitator, supports the team during planning, performance and critical situations. It is essential that the facilitator supervises the team- coach/client cooperation during the project's duration.

5. **The real time projects should act as an evaluation of the program success.** While focusing the project target on a specific short-term goal, the team should provide a written account of all goals agreed on. During the different working phases the team should be obliged to issue monthly reports. A final result report, after the completion of the project, is essential to quantify and qualify the gained results in comparison with the primary targets. This report should contain propositions for a general implementation of the concrete results in order to make valid data available inexpensively and to integrate the evaluation in the working process. The evaluated data provides feedback to every group involved: participants, coach and facilitator.

6. **The gained experiences should be shared within the whole company.** Online archiving of every running and completed project makes experience and project data accessible to every employee. Widespread information on various topics and problems can lead to an active best practice and experience sharing. Members of the management learning program should have an additional access to private knowledge pages where further information on competencies, skills and special subjects of each former and present participant can be shared. This leads to a pool of experienced and competent contact persons, accessible to everybody involved in the management learning projects.

Discussion Questions

1. What lead to the founding of Siemens Management Learning in 1997?

2. Which innovations did the Siemens Management Learning Programs introduce?

3. Which challenges do the participants of a Business Impact Project have to face?

4. How does not only Siemens but every participant benefit from a successful BIP?

5. What future challenges for the Siemens Management Learning Program do you see and how could these be tackled?

E-Learning and Knowledge Management, symptoms of a new Siemens reality

"Siemens Learning Valley Belux"

Jef Staes

Abstract

A growing number of companies and organizations are developing a new learning culture, since they realize that learning faster than the competition is the only competitive advantage which is viable. Those who learn faster will be capable of renewing their products and services faster. They will have the advantage in the current competitive market. However, getting an organization's learning capacity to gain momentum is not that easy. It is not just a question of installing new learning or Knowledge Management technologies; it requires a transformation that will impact the whole organization. How do you do this and how do you sell your approach to an entire company?

This case study offers some thoughts on the essence of learning organizations and how e-learning and Knowledge Management can help them to become a learning organization. It places the "Siemens Learning Valley" for Belgium and Luxembourg in the right perspective. It's a story about the birth, the vision, the strategy, the results, the challenges and the future of Siemens Learning Valley as a project that drives the learning power of the organization.

The beginning of a journey

In 1998 different events triggered the need for and the conception of the Siemens Learning Valley. First there was the call from the CEO of Siemens Belgium and Luxembourg (Belux), Francis Verheughe, to find a way of integrating the activities of the different business units' training centers. How could we arrive at a common and more efficient training offer for our employees? How could e-learning help us to achieve this? His second request was about Knowledge Management. What did Knowledge Management mean to us and how could it be implemented within Siemens Belux? A project was initiated that aimed to test knowledge sharing and the knowledge management process. Some months later a Business Impact Project (BIP S3 Management

Learning) was started with Francis Verheughe as sponsor and Arnold De Decker as coach. This BIP was of great assistance in helping the Siemens Learning Valley gain momentum within Siemens Belux. The conclusive event was the completion in 1999 of a Dutch book by Jef Staes: "*Het herexamen van een managementgeneratie*". In English this would translate to: "*A management generation's last chance*". The book contributed new insights into and concepts about learning and the importance of learning in the New Economy. Some of these new concepts were used to develop the Siemens Learning Valley.

The projects were carried out during the period 1998-1999, discussions were held between the different members of the project team and learning took place. New learning and knowledge management applications were also piloted. At this stage the name "Siemens Learning Valley" did not as yet exist for the various projects. Different consultants from inside (Siemens Business Solutions or SBS) and outside Siemens were brought aboard to contribute to the discussions.

One of the main discussion points was the meaning and place of Knowledge Management in relation to training and e-learning. Nobody really had an answer to these questions and finding one seemed impossible, since those responsible for Knowledge Management and those responsible for training and e-learning occupied conflicting positions. We nevertheless believed that conflicting views and conflicting points of interest could lead to the best solution. We found our solution in the magic word: *learning* – we had to focus on the *learning activity* of the employees. From this perspective, Knowledge Management, knowledge sharing, training and e-learning are solutions for learning. Once this common understanding had been arrived at, the integrated project could truly start. The name of the project became the "Siemens Learning Valley – Where Knowledge Shapes (Y)our Future". This common understanding gave us the opportunity to develop a total solution for Siemens Belux, a solution that had to support the growth to a learning organization. It had to be a solution where training, e-learning, Knowledge Management and sharing – all sides of the same coin – were used as a leverage for learning.

During June 1999 the project was explained and demonstrated to the top management of Siemens Belux as an application of the Siemens Belux mission, vision and objectives, derived from the Siemens values and Top$^+$ program and integrated into the corporate Balanced Scorecard. The goal of this event was to obtain support from the different business units and find the necessary commitment to carry on and finance the journey. We were successful: both the venture capital and the mandate to explore the unexplored were ours.

During 2000 the first Siemens Learning Valley (SLV) portal was developed and the first knowledge management, e-learning and training applications were developed or acquired. The rollout of SLV within Siemens Belux was introduced by means of a "road show". Different Siemens locations were visited during which the story of the Siemens Learning Valley was told. Siemens employees also had the opportunity for hands-on experience with the different applications.

At the time of going to print the journey continues, although it is never easy. We face new challenges every day. We nevertheless realize that these challenges contribute to the maturation of the Siemens Learning Valley, while the vision and strategy behind the Siemens Learning Valley keep us on track. Moreover, these challenges tell us where to build new tracks. The year 2001-2002 will see a further integration of SLV in the daily business of Siemens Belux – not because we want it, but because it is needed.

This year will also be the year we extend our cooperation with Siemens headquarters. It's very important that Siemens Learning Valley doesn't become an isolated project within the Siemens global community. For this reason the Square (see later), one of the key applications in SLV, and ICN ShareNet will be combined into a new Siemens Sharenet. New, innovative features will be added which will even better support the exchange of knowledge within the organization. The Siemens Learning Valley is committed to this development and will share its knowledge when developing this new environment.

Today, at the start of its journey, the Siemens Learning Valley has already been cited as a best practice inside and outside Siemens, not because the project has been completed, but because we are pioneering in the field of learning organizations. More background information about the Siemens Learning Valley follows below.

The conception and birth of a new economy

All over the world motivated people are developing their competencies by using information as building material. Their brain is a never-ending construction site where old knowledge, skills and attitudes are replaced by new paradigms. These paradigms make it easier for them to conceive new products and services, some of which, backed by investors who believe in them, find their way to the market. These new products and services feed the market with new information. Then the cycle restarts. This *circle of innovation* is a natural process in learning organizations.

It is not difficult to understand that this process is responsible for an exponential growth of information. Take the rapid deployment of information technology and worldwide telecommunication infrastructures into account, and you can imagine that the process depicted in Figure 52 is speeding up. If you as an organization or an individual, cannot increase your learning power (development of new skills, knowledge and attitudes), or your power to change (use resources to transform your business processes based on what is learned), then you will have to leave the market.

It is therefore imperative to become a learning organization. Peter Senge describes learning organizations as follows:

"The real competitive advantage for organizations is their ability to learn faster than the competition, to generate and share knowledge (learning power) and to continuously improve (power to change). A learning organization is an organization that has the capacity to build her own future".

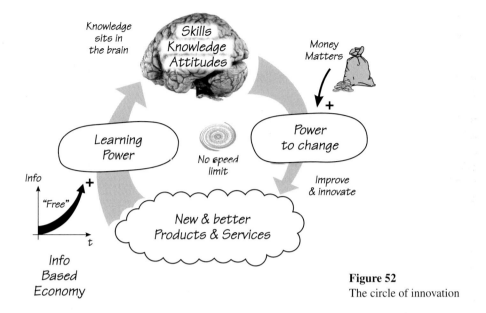

Figure 52
The circle of innovation

The new global reality

The new global reality is based on the existence of a new, "fifth element": *information* (Figure 53). Besides the four earthly elements: water, air, fire and earth, we now have a fifth element. This element is not produced by our earth, but by us, the human resources. Without us, information wouldn't exist the way it exists. We, the learners and the innovators among us, are creating a new global reality based on information.

The entire economy and society will henceforth be influenced by this new element. It's the driving element of the circle of innovation. This will, no doubt, have significant consequences for our manner of learning, using, accumulating, generating and

Figure 53
Information as
the 5th element in
a new global reality

243

exchanging knowledge. We have to adapt to this new environment which demands much from people, many of whom still regard information as a threat. It is not a threat; it's here to stay and we have to adapt ourselves to this new situation.

The exponential growth of information has an enormous impact on at least two kinds of organizations.

- The first kind is the *educational institute*. It is no longer possible to adapt our standard classroom courses to keep up with the information and the new knowledge that are becoming available. Classroom teaching can't keep up with reality. We have to look differently at the role of the educational system of a dramatically changing society. The first priority is no longer to fill students' brains with knowledge, but to teach students "how to breathe information". Without this skill it is going to be impossible to survive in an ever-changing world. To achieve this transformation, teachers have to embrace the new society and look for a new role for the educational institute. Without this commitment the educational system will lose its driving force in the new global reality.

- The second kind is all *companies* competing with one another in the free market economy. For them information and the processing of information (learning) represent the only way to survive in the new global reality. If management doesn't succeed in developing a new management style that promotes learning and competence development, then organizations will cease to exist. It is this certainty that causes much of the stress in organizations. The road to this new culture is not simple and will demand much self-sacrifice from many managers.

Learning power, a process captured in *Jeff's Law*

Learning power is a person's capacity to capture and use information in an effort to develop new skills, knowledge and attitudes. Without sufficient learning power, people aren't able to adapt swiftly enough to what is happening in the world. Without this power the circle of innovation is fiction.

The *learning tension* is almost a physical tension in our brains. The stronger the tension, the larger the hunger for information. People in the right roles develop a high learning tension. Then information doesn't have to be pushed towards employees, because employees are motivated enough to pull the appropriate information from the information flow.

The *learning skill* is represented by the angle in the picture. You have a large learning skill if you are able to find information in the information flow. This means, for instance, that you know how to use a search engine on the Internet, that you are able to participate in online Communities, that you have excellent social skills (i.e. assertiveness) and are able to work as a team member. If you don't develop your learning skills you may have a high learning tension, but you won't be able to find the right information and your learning power will be small.

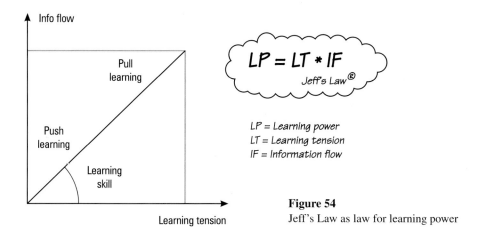

Figure 54
Jeff's Law as law for learning power

At first glance it appears as if this challenge, enlargement of the learning power, will be too difficult for many of us. This way of thinking – this paradigm – is a result of the schooling we received. We subsequently regard learning as something imposed on us: A teacher explains, pushes information towards us, and an exam follows to check if the knowledge has been captured. This mental model of how learning occurs is no longer appropriate in the New Economy.

People will only plug into the information flow if they are motivated to learn, in other words, if they have a job or a role in an organization that they like. Without this intrinsic motivation, or learning tension, people will not learn swiftly enough to keep up with change. We could even go as far as to say that the "treatment" of the human resources in an organization will determine the speed at which that organization learns and changes. This new "people management" will be the most important incentive for learning organizations.

To capture all this in a law, I depict learning power as the product of learning tension multiplied by the amount of information somebody uses to develop his competencies (Jeff's Law).

Learning Power = Learning Tension × Information Flow

This analogy is based on the basic law of electricity where the electrical power equals the electrical tension multiplied by the electrical current.

If you don't want people to learn, remember the following statements:

- People don't learn if you place them in the wrong role, or if you don't give them access to information (learning tension = low, possible information flow = low). People with professional ambitions will move to another organization or team if you can't offer them challenging jobs and learning opportunities. People without ambitions will stay.

- People won't learn if you put them in the wrong role, although you give them access to all the information you have (learning tension = low, possible information flow = high). This statement is very important, because in this case investments, in for instance e-learning and Knowledge Management, will not result in a higher learning power. Communities of Practice will be initiated, but will not last because they miss intrinsic motivation.

- People won't learn if you put them in the right role, but don't give them access to information (learning tension = high, possible information flow = low), such as refusing them access to the Internet, or don't invest in e-learning systems or Knowledge Management. It is very possible that an organization will lose these driven people, because they will look for another environment in which they can grow.

It is therefore very important to realize that e-learning and Knowledge Management or sharing will only happen in an organization where people can migrate to challenging roles in which they can give the best of themselves, and where managers know how to motivate and mobilize people for the achievement of compelling visions and strategies. E-learning and KM are systems that will only be used by highly motivated employees. These systems are pull-learning systems. You can't introduce knowledge management and e-learning systems without changing the whole culture of your organization. E-learning and knowledge sharing are symptoms of learning organizations; they don't make learning organizations, they support them.

The three steps towards e-learning and knowledge sharing

All that is missing from Figure 55, is the importance of the three steps. Miss one and you fail. This is true for many organizations that have failed in their change project.

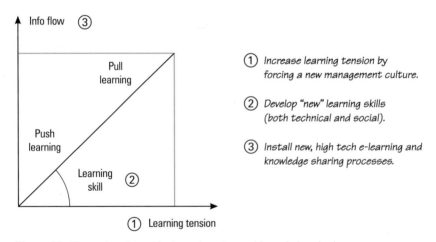

Figure 55   Three steps for achieving e-learning and knowledge sharing

The first step, changing the management culture, is the most difficult step to take. Once this step has been taken the other two will follow spontaneously. E-learning systems will be installed and Knowledge Management or sharing will occur, because technology is really not the issue. Most of the systems can be bought or developed, and the skills needed to use these systems will be developed/trained while people use them.

The Siemens Learning Valley

When installing the Siemens Learning Valley in Belgium/Luxembourg, the most important point was most probably the creation of a vision and a learning strategy. Without the vision, explained elsewhere, the learning and KM applications that we bought or developed would just have been technology. The vision leverages the learning organization.

The valley metaphor and logo (Figure 56) evoke the image of water running through the valley. That water is the knowledge flow which nurtures fertility, allowing new ideas, products and services to emerge. Everybody must realize that they can learn from the knowledge flow and can contribute to it.

The green fields in the valley represent the employees. In an organization they have to be in the right place to grow. The same is true in nature where every plant has its own specific place where it flourishes.

Figure 56 The Siemens Learning Valley logo and metaphor

The sun in the valley represents the mission and the vision of the organization. Employees have to develop their competencies to contribute to the achievement of the vision. The same is true in nature where plants grow towards the sun.

Decide for yourself...

Employees now have the freedom to choose their path of learning, but are, simultaneously, also responsible for it. However, this choice is only possible if e-learning and KM applications are at their disposal, enabling them to embark on their paths of learning and the acquisition of knowledge in a flexible manner. The Siemens Learning Valley does not merely offer the freedom to choose **what** to learn, but also **when** and **how**. All of us can decide on our learning projects according to our workload and timetable. We learn when we want, not when we have to.

The Siemens Learning Valley Portal

Fortunately, Siemens has practically all the technologies to create all of these learning applications on PC, on the intranet, the Internet and elsewhere. It will probably be more difficult to cultivate a new attitude. Indeed, the training programs will from now on no longer be served as one whole menu. As Francis Verheughe, CEO of Siemens Belgium, says:

"Staff will have to choose à la carte. They cannot sit back and wait passively until specific courses are offered, but need to go hunting for the information that will increase their competencies. They are in the middle of the valley and must know which information sources and knowledge streams they should drink from to flourish and bloom".

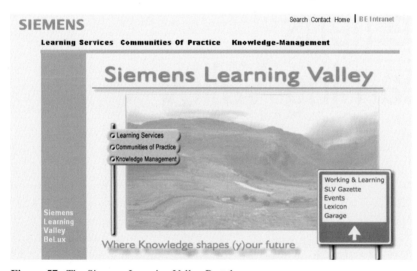

Figure 57 The Siemens Learning Valley Portal

To make all of this possible we created a learning portal that allows all people within Siemens Belgium and Luxembourg and even worldwide, to learn and share their knowledge with others. It is the entry point to a whole new world of learning – http:// intranet.slv.siemens.be (Figure 57)

With the Siemens Learning Valley Belux we are creating an environment where people with a high learning tension will find an answer to their hunger for information. This is one of the essential steps on our way to becoming a learning organization.

"AAA Learning" Strategy

Our solution is not just a selection of technologies or applications, it is based on a challenging vision of and strategy for learning. Our AAA-learning strategy (triple-A learning Figure 58) has been the driver towards the application selection. AAA-learning combines training (classroom), e-learning and knowledge sharing by means of Communities of Practice. Keeping this concept in mind, it was clear that systems currently available on the market are not always suitable for the learning model of the future. Learning, coaching, content creation etc. can no longer be centralized, the *networking* and *sharing* of knowledge are going to be the answer to tomorrow's competence development. We therefore built our own learning environment and applications.

AAA learning is based on Jeff's Law. We are convinced that the learning behavior of people changes when their learning tension increases.

Awakening
If the learning tension is low, people will "only" learn in a classroom environment. Information has to be pushed towards the learner. This method of learning is still acceptable, but in our concept it will be used to awaken people. This learning behavior alone does not enable one to keep up with the New Economy.

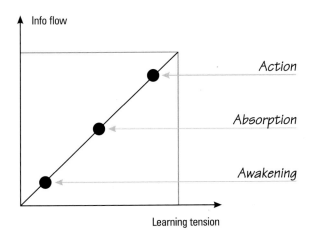

Figure 58
AAA Learning

Absorption

If the learning tension increases, people's learning behavior will change dramatically. They will not only learn through classroom training, but they will start using other learning systems. It is in this phase that e-learning, or web-based learning will start occurring.

Action

Action learning is the learning behavior seen in people with a very high learning tension. They will start participating in *Communities of Practice* where they will share, for instance, best practices, not because they have to, but because they understand that if they really want to become an expert they have to share information and knowledge with others. This is the "GATA" age. An age where people "Give Away" and "Take Away" information to develop their competencies.

Blended learning

A typical learning trajectory could look like this: people are invited to awakening sessions. During these, standard classroom events, a coach or trainer gives sufficient information and knowledge to awaken these people. They have to become consciously aware of their own incompetence. If their learning tension increases (they have the right role in the organization), a different learning behavior develops. They will be much more motivated to learn via the web, go through e-learning packages (read a book), or attend a more intensive classroom event. During this phase they absorb a lot of information, and this phase is therefore also called the absorption phase. Really driven people don't stop their learning process. They are so motivated to develop their competencies that they start with action learning. During this phase they start with knowledge sharing in Communities of Practice.

This learning process, in which different learning environments are used, is called blended learning. With AAA learning we look at learning, or competence development from a holistic point of view. The different parts are not important, but we should focus on the whole process.

It's very important for an organization to reengineer its learning process so that people with different learning tensions can learn in different learning processes. To this end the Siemens Learning Valley installed different applications.

Communities of Practice

A Community of Practice (CoP) is a flexible group of professionals, informally bound by a class of common interests, and the pursuit of common solutions, and thereby embodying a store of common knowledge. The main purpose of a CoP is the exchange of knowledge between its members.

During this process different players are active in a process of knowledge sharing. In the AAA learning strategy this supports the action learning process. The different players and activities in Communities of Practice can be seen in the Figure 59.

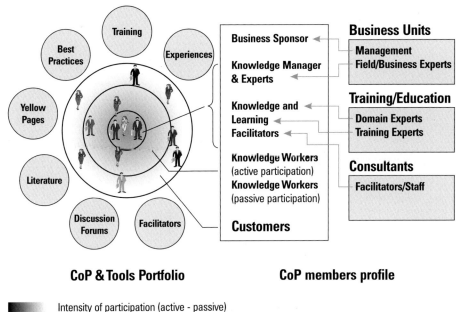

CoP & Tools Portfolio   **CoP members profile**

▆ Intensity of participation (active - passive)

Figure 59   Activities and players in Communities of Practice

Siemens Learning Valley Applications

Training Center Manager or TC manager

This is our Learning Management System for standard classroom courses. TC Manager (Figure 60) can handle all tasks related to a training center and has web-publishing facilities that support online shopping (module creation, event creation, online enrolments, reporting, statistics, communication, invoicing, multiple catalogue support, client relationship management, reservation of trainers & rooms etc.). TC Manager is furthermore able to interface with other systems, e.g. Peoplesoft and SAP.

Employees are able to browse through the training offer of the different training centers within the business units and are able to register online.

Some statistics:
Number of course days organized by TC manager on yearly basis: 5,000
Number of student days organized by TC manager on yearly basis: 30,000

Horizonlive as a virtual class

The Siemens Learning Valley (SLV) has installed Horizonlive to offer online classroom learning (Figure 61). This application is used to broadcast and capture live classes via the intranet of Siemens Belux. Its archiving options and stand-alone creation module,

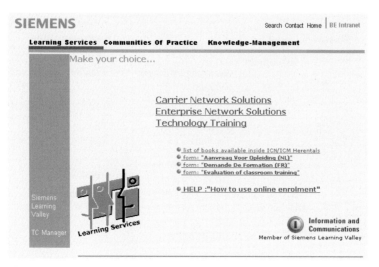

Figure 60 TC manager

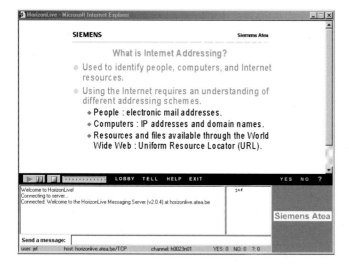

Figure 61
The Virtual Class
in action

together with its fully web-based access make HorizonLive a powerful tool for learning. Interaction is possible via chat-rooms and telephone. Break-out rooms, hand raising, participants list etc. are at your disposal. Applications sharing and white boarding are standard features these days. All in all, an easy to use learning environment.

Archived sessions are added to The Square (see later). At present this technology is no longer new, but it still has some technical drawbacks. Siemens Learning Valley Belux is experimenting with this technology. The year 2001-2002 will be used to further deploy this technology within the organization.

Number of archived sessions in the square : 140

The Square

The Square is a tool that was developed by Siemens Belgium and is a very powerful tool for web-based learning and Communities of Practice (Figure 62). It is a collaborative platform to leverage the exchange of knowledge.

The 100% web-based environment is based on a pool of domains (learning topics, learning communities, Communities of Practice) and per domain a domain manager can activate several services (e.g. information, news, mailing list, discussion forum, coaches, sources etc.). The domain manager can control the access rules per service per user group. Creation of domains, user groups, users etc. is very easy; within 2 minutes a new domain can be set-up with all services activated.

The major differences between The Square and other learning platforms are: its distributed responsibility concept, its facilitating of publishing and its control of quality. Moreover, The Square offers you extended personalization functionality on the domain level and modules, as well as via the My Square and My Profile functions. The Square helps you to *give away* and *take away* knowledge, but especially to build networks with your peers throughout the world. Whereas the Siemens Learning Valley started as a Belgium/Luxembourg initiative, more then 35% of the current users are from more than 50 countries and from different business units.

Some statistics (status June 2001):
Number of user logons : 61,000
Total number of registrations : 3,858
Number of Belux registrations : 2,528
Number of international registrations : 1,330 from more than 50 countries worldwide.
Number of communities and sub-communities : 30 very active communities and more than 100 less active (starting communities, action teams etc.). Some of the communities are international.

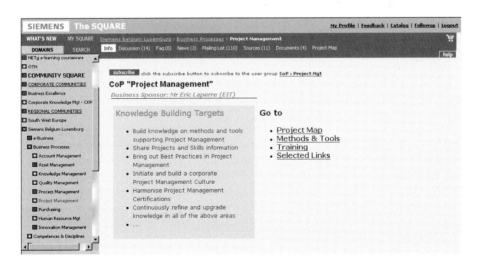

Figure 62 The Square

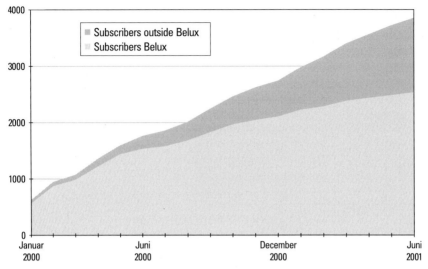

Figure 63 The evolution of new subscribers

Number of discussions : 2,400
Number of sources and documents : 4,800

The evolution of new subscribes can be seen from Figure 63.

KLIX (Knowledge and Learning Index)s

In future the SLV will be able to measure the evolution towards a learning organization as well as present an index that illustrates this. It will also measure the increase in Human Capital in the organization. We believe that it will become increasingly difficult to calculate the return on investment in learning for every learning event. The list of learning events is long: classroom training, meetings, e-learning, on-the-job training, Communities of Practices, reading of books and other literature, learning from the Internet, from colleges, from giving presentations, from mistakes, from best practice sharing etc.

It will not only become very difficult to calculate the cost, but also the calculation of the direct impact on business results will be difficult. And yet, a learning organization is an organization that embraces learning in all its forms.

KLIX, a project that is currently in the pilot phase, will provide a new approach to "calculating" the results of learning by monitoring a selection of investments made in learning (indices) and then relating this to corporate results. It is a pioneering approach to link learning to business results. It is our opinion that by exploring the measurement of fluctuation in Human Capital (part of the intangible assets) we can foresee corporate results.

Siemens Learning Valley Gazette

Communication is one of the most important issues when developing a learning environment. To this end we started a publication called the *SLV Gazette*. Currently it is published once every two months and is distributed to the 1000 employees who have subscribed. Initially we didn't push this information channel, since the SLV still had to grow. As we are now convinced of the maturity of the Siemens Learning Valley, we intend to promote this approach more aggressively. In future the SLV Gazette will be published every month and distributed to all employees.

Some general remarks about the applications

- The development of the SLV solution is the result of a *common vision and strategy*. This means that all the applications and functionalities contribute to the same process.
- The applications support the processes for Business and People Transformation (culture change, new learning and new management culture).
- An open architecture makes it possible to integrate several applications in order to build a global and complete solution.
- The added value of an integrated platform for training, Learning Communities & Knowledge Sharing Communities was a key-success factor.
- The solution allows the user a robust, fast and cost-effective implementation.
- The solution is user-friendly, allows delegation of responsibilities, promotes decentralized administration and self-management possibilities for the user (flexible and cost effective).
- Integrated and dynamic competence profiles of the users of the applications are automatically generated.

Siemens Learning Valley business unit

The Siemens Learning Valley business unit, with Karel Verhelst as the business unit manager, was created in April 2001 to commercialize the concepts and applications that drive SLV Belux. This strategic move was taken to valorize the knowledge and experience within Siemens Learning Valley Belux on a market where there is a growing potential for products and services related to e-learning and Knowledge Management. Our holistic approach towards learning differentiates us from other vendors on the market.

Our first customers have already found their way to Siemens Learning Valley and more are requesting information about our portfolio of products and services. *The following external customers already work with SLV : The Ministry of Public Affaires in Belgium, The Ministry of Education in Luxembourg, Dutch PTT in the Netherlands, Fujitsu Siemens Computers in Germany.*

To clarify matters for the reader: Siemens Learning Valley Business Unit is a Siemens business unit founded in 2001. Siemens Learning Valley Belux has become a project in the new business unit, the customer for this project is Siemens Belgium & Luxembourg.

Conclusion

It's important to realize that e-learning and Knowledge Management will dramatically change the face of learning. This will not be the case for everybody, but most certainly for those who work or live in a challenging environment that stimulates the development of new skills, knowledge and attitudes. The impact that e-learning and Knowledge Management or knowledge sharing will have on the learning power of individuals and organizations, will lead to more dramatic differences between individuals and organizations that take part in this new global economy and those who don't. The choice is yours.

The Siemens Learning Valley Belux has been operational since the beginning of 2000 and the feedback we are getting is very encouraging. We see a daily increase in users, nationalities and knowledge added to the system. All applications and the concepts behind the Siemens Learning Valley are based on the mental image explained in this article. We feel we struck gold with our approach to e-learning and Knowledge Management and are facing a tremendous opportunity with this concept. We have to keep on learning and investing to keep the momentum going. Currently we are looking for international partners who share the same ideas about knowledge creation and sharing to further develop the Siemens Learning Valley. Feel free to contact us (e-mail: slv.helpdesk@siemens.be).

Feel free to learn and to share.

Key propositions.

1. **Conversion between KM, e-learning and training.** Knowledge Management, knowledge sharing, training and e-learning aren't different solutions to different questions. They are part of the same solution to one question. How can we make sure that the organization learns faster? It's very important to ensure that while initiatives are taken to introduce Knowledge Management, training departments also reengineer their training solutions. In the end this means they will have to migrate to e-learning and knowledge sharing (Communities of Practice). This conversion is already taking place in the world around us. Siemens Learning Valley supports this conversion.

2. **Use the "training centers" as a leverage.** The conversion taking place makes it very important that Knowledge Management is not seen as a "stand alone" solution. We also have to use the enormous potential and power of the training departments as a leverage to introduce not only e-learning, but also knowledge sharing (Communities of Practice). We have to refocus the existing potential. One of the major issues within Siemens Learning Valley Belux is streamlining the different training centers in the different business units. This demands a global approach to the issues at hand.

3. **Competence gap.** There is a growing competence gap between the practitioners of e-learning and Knowledge Management, and the organization itself. We have to realize that not only employees, but also people in HR and training departments have to develop new competencies to use e-learning and Knowledge Management in their daily practice. They have to become missionaries in the new world. Today they lack the needed skills, attitudes and knowledge to take full advantage of the possibilities already offered. It's a challenge that should have first priority. E-learning and Knowledge Management should be integrated into the daily business by those people already supporting the daily business. We have to use them as leverage for change.

4. **Management culture.** We have to realize that when introducing new ways of learning through e-learning or Communities of Practice, we introduce a new learning culture. Employees need to be empowered to learn when they need to develop their competencies. Not the manager, but the employee himself will be responsible for his learning. These kinds of projects therefore need management support, not only because of the venture capital invested, but also because of the culture change that is caused by the new ways of learning. Management has to realize the impact thereof on the culture. A new management culture is mandatory.

5. **You need time.** You don't become a learning organization by installing new learning (knowledge sharing) applications, you become a learning organization by increasing the learning tension of all your employees. This is the real challenge for management. E-learning and Knowledge Management are symptoms of a new management style. Therefore developing a really flourishing Siemens Learning Valley demands time to develop. It is not the tools, but the people who make the difference.

6. **Return on investment.** It's my personal opinion that calculating the return on investment for investments made in learning or Knowledge Management will be nearly impossible. It will be very difficult to trace the learning time (cost) invested by the employees and it will be even harder to trace those learning events that resulted in a certain profit. Learning in the New Economy has to be promoted and supported by believing in it. You have to believe that the investments in learning and Knowledge Management support the achievement of your business goals. Perhaps this "belief" will be the most difficult management paradigm to change. It's similar to believing in the power that results from developing a compelling vision for your organization. How do you calculate the return on that investment? New ways of relating learning effort to business results should be explored. KLIX is one of these experiments.

7. **Casting for a vision.** Middle management plays a very important role in the story of a learning organization. It has to develop a compelling vision for every team and has to ensure that all employees can migrate to their right role. This is one of the most important responsibilities of management in a learning organization. Without this vision and casting of employees, the learning power of the organization will be too low and the learning power will not be focused enough to achieve the vision. While Siemens Learning Valley develops its applications, middle management has to acquire the necessary skills and attitudes. HR plays a very important role in this chapter.

8. **Learning Portal and application development.** After a year of being online, Siemens Learning Valley has realized that we have to continuously invest in the different applications. The learning process we, and the users, are experiencing gives us a better insight into the real need of the organization. It's from this understanding that we will develop a new portal and continuously improve our different applications. The Valley is a living organism and it needs continuous attention.

9. **Siemens Learning Valley as a metaphor throughout the world?** During the past year the Siemens Learning Valley has attracted much attention from both inside and outside Siemens. Perhaps it is possible for us to realize a dream in which all Siemens employees go to the same Siemens Learning Valley and learn what they have never learned before. Perhaps we can create a Siemens Learning Valley for our customers, a single entry point for all their learning needs, perhaps…

Discussion Questions

1. Discuss the different events that triggered the need for and the conception of Siemens Learning Valley Belux. Do you see the same need in your organization?

2. Knowledge sharing, training and e-learning are just other ways of learning. Do you agree with this statement? Why? Why not?

3. The circle of innovation is a natural process in learning organizations. Explain the different elements in this circle of innovation. How does it explain the exponential growth of information, the 5the element?

4. Explain Jeff's Law. How can it be used as a holistic approach towards learning? What is the meaning of learning tension and how does it relate towards AAA-learning? Discuss the meaning of Jeff's Law in relation to knowledge management.

5. Three steps are needed to create the learning organization: development of a new management culture, development of learning skills and the development of new learning processes. Why do you need all three of them simultaneously? What happens if one of the steps is missed?

6. Siemens Learning Valley Belux deployed a few learning applications. Explain the purpose of every application. Why do we need a global approach for Siemens to deploy these kinds of applications?

VI Visualizing more of the Value Creation

Getting real about knowledge sharing: the Premium-on-Top bonus System

Michael Gibbert, Petra Kugler & Sven Völpel
prepared this case under the supervision of Professor Marius Leibold.
Premium-on-Top is based on a model by Nicole Prummer.

Abstract

During the last decade, Siemens' Information and Communication Networks department (ICN) was confronted with fundamental changes in its environment. The new situation required the manufacturer to newly define its business which had shifted from a stable and simple product business to complex customer-oriented services. In describing the close collaboration between the South African key account manager, Doug Williams, and the German project manager, Markus Schmid, the case shows the process through which Siemens ICN laid the groundwork for a new business model based on intensive global collaboration and Knowledge Management. The basic idea was to make knowledge, that had been created somewhere in the world for ICN, available for reuse and innovation elsewhere. To give explicit and tacit knowledge sharing a nudge, two projects were combined. On the one hand there was the Knowledge Network Model (LITMUS) project, through which international Knowledge Management is measured, and on the other hand there was the Bonus-on-Top, which makes worldwide knowledge transfer and creation attractive by offering employees valuable incentives. During the realization of both projects, different process stages had to be traversed: the first was the emergence of an innovative, strategic idea in a bottom-up approach, while the second was the consideration of the main aspects of a new business model. As a third step, the case clarifies the visualization of value creation through knowledge, and the motivation for active Knowledge Management.

Introduction

After spending two months at Siemens' headquarter in Munich, Key Account Manager Doug Williams was finally returning home to South Africa. As Doug thought about how much he was looking forward to seeing his wife, he reflected on how profitable his visit to Germany had been. During the past few weeks in Germany, he had not only consolidated his reputation as a marketing and sales expert of intelligent business-process solutions in the banking sector, but he was also returning equipped with a handful of innovative ideas that he could share with his customers.

These ideas were the products of stimulating encounters he had had with some of his German colleagues. Doug knew that if he needed advice or information from his German sales and marketing colleagues, the newly created knowledge network would be a great help. Best of all, his close collaboration with Markus Schmid, a German project manager, had developed into a good friendship. This burgeoning friendship was just one step in the intercontinental cooperation within the Siemens Information and Communication Networks (ICN) division, a global provider of telecommunication solutions.

During his 20-year career with Siemens, first as a salesman in the United Kingdom and now as a Key Account Manager, Doug had gathered much specialized knowledge on the market, the selling process and recently, on customized telecommunication solutions – especially in the South African banking sector. Six months ago, he had received an urgent request through ICN ShareNet, the division's worldwide, internal knowledge network (Intranet). This had been the first contact between Doug and Markus.

Markus had written:

We are presently putting together a proposal for one of Germany's largest banks for the provision of a turn key communications solution. This will involve Hicom 300E, e-trade, Unified Messaging, PCs, etc. We urgently need success stories for this type of proposal. If you can assist please let me know.

After an initial exchange of experiences via ShareNet, Markus soon realized that not only would his project group of sales and marketing colleagues profit from Doug's extensive knowledge, but that Doug was also very knowledgeable about the specific area of customized telecommunication solutions which, in turn, could be used strategically. This meant that he would have to meet his South African colleague personally. He then invited Doug to the Munich headquarters of Siemens ICN for training purposes.

A hectic time of project work and conferences, workshops and official training sessions followed. It soon became clear that not only Markus would profit from the knowledge generated in South Africa, but that the reverse was also true. Doug had learned about European customers in general and, in particular, he had learned a few successful selling strategies that had been used in the German market. The two representatives from two radically different countries recognized that the exchange of experiences not only meant that they would save time, but that their cooperation would be an impetus for the creation of additional value for their respective companies.

The crux of the matter was that knowledge sharing does not occur naturally, or by chance – especially not when it is on a global scale. Furthermore, collaboration like that between Doug and Markus poses quite a challenge. It leads to many questions that need answering by both parties, like:

- Which company is responsible for the travel and personnel expenses and/or opportunity costs?
- Who will benefit most from the arrangement?
- Which company should carry the technical and financial risk?
- Which company should receive the projected revenue?
- And, most importantly, why should the South Africans make their knowledge available to their German colleagues?

There were further challenges on the organizational, personal and cultural levels. It was, therefore, of the utmost importance that something was done to give knowledge sharing a nudge, but what?

Siemens ICN found a solution by combining two existing projects:

- *Knowledge Network Model (LITMUS)*
  through which international collaboration and knowledge exchange are measured, and
- *Bonus-on-Top*
  which makes worldwide knowledge transfer and creation attractive by offering employees valuable incentives.

The original stimulus for both these projects was twofold. On the one hand, Siemens ICN had had to contend with a drastic shift in the business environment that had turned its rather straightforward business into a demanding, knowledge-intensive task. On the other hand, the innovative idea of international collaboration, which had emerged from contact with one of Doug's South African customers, had become a strategic issue for ICN.

Let us take a closer look at these two factors that led to such radical change.

Giving knowledge sharing a nudge: a transitional environment

Siemens' Information and Communication Networks division (ICN) is a leading provider of telecommunications solutions (corporate networks, carrier or mobile networks, and Internet solutions). ICN, with revenues of 25 billion DM, operates in more than 150 countries worldwide, and has about 60 000 employees.

The company's traditional business used to be quite stable, as well as simple and straightforward. Its customers, mostly state-owned telephone companies, wanted well-defined products that could be integrated into existing networks by the customers themselves. The manufacturers were, therefore, asked to provide pre-determined products to ICN's highly specialized customers. In company terminology, this was referred to as "box selling". To guarantee a lasting business relationship, it took little more than the delivered products fulfilling the technical state-of-the-art requirements. When it came to maintaining a competitive position, quality and reliability far outweighed speed.

However, over the last decade, a shift occurred in the company environment. Times changed. Technical progress, increasing global cooperation, and deregulation of markets, among other factors, required a new type of telecommunication operator. The situation changed for both clients and manufacturers. The existing competitive structure, patterns of behavior and intensity of competition were constantly altering, making change a permanent state rather than a transitory phase. Demanding end-users required innovative and equitable network solutions and, in turn, the new telecommunication providers posed new challenges to manufacturers like Siemens ICN. These challenges ranged from finding complex conceptual packages, or designing business plans for innovative solutions to customer-oriented services. This shift, from a formerly simple product business to a new service-focus and solutions approach, required new skills and a new mindset from the manufacturers. As one Industrial Relationship Manager noted rather wryly, "We will have to unlearn thinking in packaged products and applications".

Finding solutions to these new challenges rapidly increased the complexity and knowledge intensity of Siemens' business. Coping with the challenges correctly, meant identifying the best practices quickly; sharing them on a global scale; and ascertaining that they were reused for profit in similar settings. Knowledge-intensive solution selling became an important value-adding activity, but was also an opportunity for developing a new core competence which could lead to a stronger competitive advantage.

As process and customer knowledge replaced the former product knowledge, the question was, who was able to deliver the solution and customer knowledge that were now required?

Doug Williams summarized the new challenge succinctly when he said:

Account management means intense relationship and self management. We have to gain intimate knowledge, not only about our customer and his business, but also about our own capabilities and accumulated knowledge. It requires the reuse and development of existing ideas.

ShareNet

Today, ICN ShareNet – ICN's global, knowledge-sharing Intranet – serves as an instrument to transfer and leverage existing knowledge. The basic idea is to make knowledge produced somewhere in the world, available for reuse, modification, adoption, or improvement elsewhere. Siemens' sales, marketing and business development employees on every continent have access to ShareNet. These employees are not only seen as potential frequent users of the network, but also contributors to this knowledge instrument.

ShareNet was created under the leadership of ICN Business Transformation Partners (BTP). Their role is to provide support to the ICN divisions and sales units in meeting their objectives and successfully implementing their projects. In this context, BTP's task was to establish knowledge and competency management within ICN by leveraging the common knowledge of all the ICN divisions and sales units.

ShareNet will be of key importance to the success of ICN's solution business. The company that applies existing experiences and competencies the fastest has a definite competitive edge over the other players. We need to be among the first to realize this strategic competitive advantage through efficient Knowledge Management.

Although each customer-specific solution requires finding a new way of doing business, market studies show that customers, no matter where they are located, who find themselves in a similar market phase, request similar solutions. An example of this is a customer in Finland and another in Australia who asked for comparable telecommunication solutions, in spite their location on different continents. A global actor like Siemens could turn this convergence of needs into a comparative advantage by utilizing former experiences.

Instead of continually duplicating work, global knowledge transfer and the adaptation of existing solutions to country-specific requirements, could shorten time-intensive work phases, like information procurement or marketing conceptions. Moreover, costs would not only be lowered, but the value could be increased, because additional niche markets could be addressed and/or new customers attracted. The combining of existing knowledge could also lead to innovative product solutions and the subsequent leveraging of ICN on a global scale.

But how was global collaboration to be achieved really quickly?

Building a global knowledge sharing concept

The changing business environment had had a direct influence on ICN's South African sales and marketing employees. They could no longer just rely on product knowledge. The nature of their positions, however, meant they were in a perfect position to gather information on new clients and their expectations. They could also obtain in-depth insight into their customers' way of doing business. The latter was especially important as existing customers placed orders in a relatively predictable way, but new customers had only unexpressed wishes that had to be leveraged.

Doug described the new requirements as follows:

What we need most is intimate customer knowledge, especially knowledge about the customer's economic situation. We have to make pro-active suggestions about where our customer's business might go – the field in which he could operate in future – and even question his present way of operating the business. Up to now we have only got involved in the sales process once it comes to ordering products and applications. The challenge is to start discussions much earlier. We have to assume the role of a strategy management consultant who interprets trends and designs business opportunities.

Still, by just listening to the customer the sales personnel should be able to integrate their ideas sufficiently to have a significant influence on the product design and configuration, marketing strategy, or even long-term planning.

Doug observed:

The customer is often not able to articulate his intentions and needs. It takes a lot of time, many meetings and negotiation before we are able to define a project aim and some milestones.

An example of how intensive collaboration can serve as a source of a new business model, is related below:

A year ago Doug, his key account team, an ICN Business Transformation Partners (BTP) manager and the local ShareNet manager gathered in the boardroom of a customer who required ICN's global sales and marketing knowledge. The customer – an independent, international, specialist banking group based in Pretoria – was keen to learn more about ICN's expertise in the provision of global solutions in the telecom-equipment supplier business.

At the first meeting, a level of credibility and trust were established between the two companies. The investment bank was a very loyal customer that utilized Siemens' voice systems, but had not yet considered Siemens a potential partner for integration and data solutions. The BTP team approached the bank about this, offering solution references from others in the banking industry that they had drawn from ShareNet. Through contacts made via ShareNet, the BTP team was successfully able to demonstrate that access to a world-wide network of experts could be provided via a knowledge system.

Proposals were put forward by both the customer and ICN. For the customer, this included the joint development of best-practice solutions, global access to expert knowledge in sales and marketing, and the leverage of global market knowledge. For ICN, the proposal included additional business – a complete solution, offering better use of knowledge to gain market access; the development of a reusable business model, or an international task force; and the eventual institutionalization of knowledge sharing as an integral component of the sales process.

By creating a win-win situation, a common innovative business model was developed. Doug soon realized that the combination of diverse knowledge sources led to synergies that neither party could have achieved on its own. This led to an interesting thought:

Why not use this idea of intensive collaboration both internally at Siemens ICN, as well as on a global scale?

The solution that emerged through this cooperation, could be further developed to leverage the whole division. Not only would operational work be simplified and expedited by the mutual transfer and exchange of existing knowledge, but new knowledge could also be produced jointly if the companies collaborated much earlier on in the work process. Indeed, experience with work teams had shown that groups composed of members with very different backgrounds and levels of education, deliver the most intelligent and innovative work solutions. This is particularly the case when the group task is new and unstructured. Additional value could be created for ICN in this way.

There were other benefits for ICN. Marketing and sales knowledge each comprise two categories of knowledge, namely explicit and tacit, or implicit, knowledge.

Explicit knowledge is easy to transfer using language or texts. Knowledge about markets or customers and their preferences fall into this category. Explicit knowledge can be easily codified, therefore electronic or written media, like ShareNet, serve as an adequate means of storing or communicating explicit knowledge. However, the transferred information and data need further explanation, or interpretation, to be fully understood. This is the second category of knowledge: tacit, or implicit, knowledge.

Tacit knowledge denotes knowledge that is subtler and is not easy to express in words. Michael Polanyi described it this way, "We know more than we know". Intuition, skills, or knowledge about sales processes are examples of tacit knowledge. The transfer of this kind of knowledge usually requires close personal contact and collaboration for the exchange of experiences to take place.

In the light of the above, it is clear that international collaboration could prove to be an excellent way of creating and spreading both kinds of knowledge.

Once ICN started attaching greater importance to international collaboration, it signified more than just another practical knowledge-sharing application on an operational level. It represented a modification of ICN's business practice and led the division in a new strategic direction.

A strategy had emerged from the operational level where it had intersected with a customer's needs. This fact in itself proved ICN's ability to leverage knowledge at different levels and points within the organization. However, before this idea could be put into practice, a whole lot of questions needed to be answered. This, in turn, resulted in the formulation of a new internal business model.

The main aspects of a common business model

The objective of an ICN-wide business model was to explain the context of and inducement for global collaboration, to share knowledge and to propose alternatives, in order to accelerate knowledge sharing proactively.

Independent, local companies, or business units needed to see the advantages of knowledge sharing to motivate them to participate in international ventures. Motivation, therefore, had to be a key component in a common business model. It soon became clear that such a model had to take several factors into account in order to identify a company's intention to cooperate as a dependent variable.

Three main groups of interdependent factors were identified (see Figure 64). All three groups take note of contextual factors (external and internal) of the individual companies and the connection between them. Consequently, the business model had to be simplified in order to set optimal global incentives for the companies that would be cooperating internationally. For example, if a local company wanted to maximize its profits, but had a low wage level and a high level of expertise, it would be easy to motivate the company to assign resources in return for money. If, however, the local company was mostly independent, had a high level of expertise and a short supply of manpower on its local markets, it would obviously be more difficult to motivate the company to support a second one with its expertise.

A brief description of the three groups of influencing factors follows.

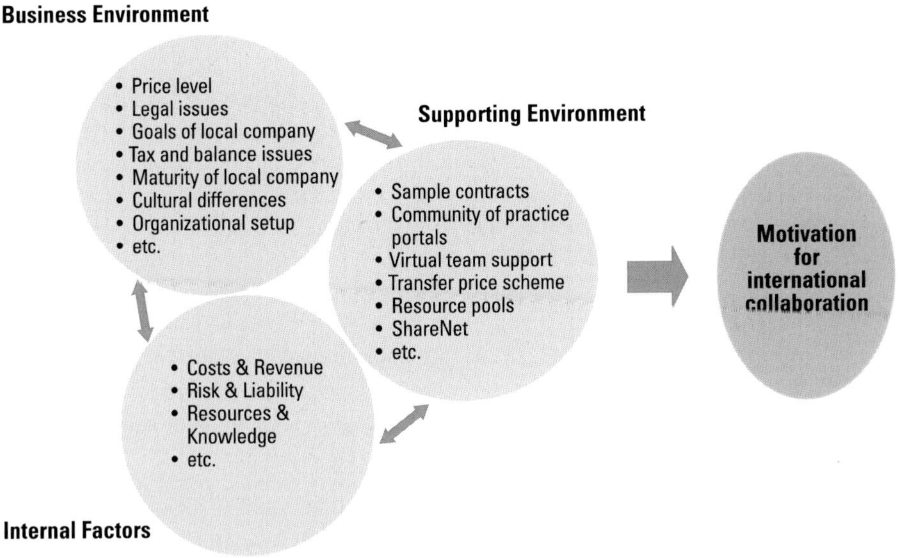

Business Environment

- Price level
- Legal issues
- Goals of local company
- Tax and balance issues
- Maturity of local company
- Cultural differences
- Organizational setup
- etc.

Supporting Environment

- Sample contracts
- Community of practice portals
- Virtual team support
- Transfer price scheme
- Resource pools
- ShareNet
- etc.

Motivation for international collaboration

- Costs & Revenue
- Risk & Liability
- Resources & Knowledge
- etc.

Internal Factors

Figure 64 Value Sharing Environment

The first group of factors constitutes the general business environment. This environment determines the external business parameters that need to be considered for the purpose of collaboration. These factors reflect the general contextual frame, for instance: cultural differences, legal issues, the overall price level, or strategic goals of the participating local companies. As a whole, this group of influences tends to be a given and has a long-term effect. As far as ICN was concerned, this group cannot be easily manipulated, or adapted in the short-term, if at all. For example, it would be very difficult to motivate a company striving for financial growth in the short-term, to accept a second local company that is striving for collaboration in order to achieve success in the long-term, as their goals are conflicting. If there were common goals, however, collaboration would leverage both companies, meaning that they might be willing to work on a common project.

A second group of influencing factors is found in the internal supporting environment of each party. These factors center around the organizational structures and tools available to local companies on a global scale. These are factors like the different kinds of team support, Communities of Practice portals, or resource pools that support the local companies in different ways in order for them to collaborate. In contrast to the first group of factors, these may be partly open to influence.

The internal parameters of the individual parties form the third group of influential factors and this group is the most important one that affects a company's motivation to collaborate. These include aspects like the availability of scarce resources, especially expertise, manpower, and revenue. Put more concretely, these factors could be things such as the costs arising directly from the common venture in comparison to the resultant revenues (for example, the costs of the consultation of the knowledge-provider).

A lack of expertise needed for a project could certainly activate a company to look for collaborating partners. The distribution of technical risks, financial risks, and liability for a specific project, is another factor that must be taken into account.

Together, the three factor groups give the overall context within which the companies operate, and which affects their motivation to collaborate or not. As the factors mentioned in the last group are the most likely to be influenced by local companies, it follows that a common business model has to concentrate on the internal parameters of costs and revenue, and risk and liability, as well as resources and knowledge. On the whole, the more a local company expects to profit from collaboration, either in terms of financial revenue or intellectual capital, the greater its motivation to contribute is likely to be. However, motivating factors for seeking international cooperation are only partly responsible for the cooperation between a knowledge-providing expert country and a knowledge-seeking country. Their motivation is usually more mutual complementary benefit.

Examples of motivation for collaboration are well illustrated in Doug and Markus' intercontinental project, which aimed to transfer the South African's knowledge to help the Germans find a solution to a turnkey communication project in the banking sector.

In this instance, the main reason for ICN Germany's collaboration was to fill their knowledge gap regarding specialized banking and to find a customized solution acceptable to their customer. They requested support from their experienced South African colleague to enable them to seize this opportunity. In addition, the initial knowledge transfer enabled the German team to accept other similar opportunities and therefore penetrate the banking market further. This collaboration led directly to negotiations for a comparable project.

During the course of the cooperation, Markus realized that his team was working day and night. Additional manpower was needed, if they were to complete the project successfully and on time. Where could suitably experienced manpower be found at such short notice? Apart from Doug, Markus had to "borrow" three British co-workers for a four-week period as an international support team for the project. This meant additional costs for ICN Germany (transport costs between the UK and Munich being one of the costs), but the overall benefit of the collaboration was definitely positive.

For the British co-workers, on the other hand, working in Germany represented not only an opportunity to collect international work experience, but also to expand their knowledge and skills. In addition, as representatives of their UK Company, they confirmed its status as a Center of Competence for complex telecommunications solutions in the banking industry.

Although the knowledge-seeking country (Germany) already had an enormous knowledge base, all three countries (the other two being South Africa and the UK) learnt something from the exercise. The exchange of knowledge between members of the global task force, with their complementing experiences and competencies, meant they could develop innovative solutions together.

Bonus-on-Top: from visualizing towards enabling knowledge creation

Building a common business model was not sufficient. An approach needed to be developed for the conceptualizing of the flow of knowledge, before knowledge could be shared between the two partners. More importantly, such a stream of knowledge has fundamental implications for value creation and these implications have to be spelled out.

A trajectory of international collaborative value creation

As a starting point for the visualizing of the value process, a trajectory of international value creation was identified, taking into account the factors that would have a direct influence on this: motivation, costs, knowledge, the customer and revenue. Additionally, three main phases, which emphasize different levels of the value-creation process, were examined more closely.

Project value

Value is first created in the context of the actual project's confines. Through the re-use of knowledge, the value-creation process should be extended to enable future returns for the business. Eventually, knowledge recombination will lead to the value being incremented through the securing of new business (see Figure 65).

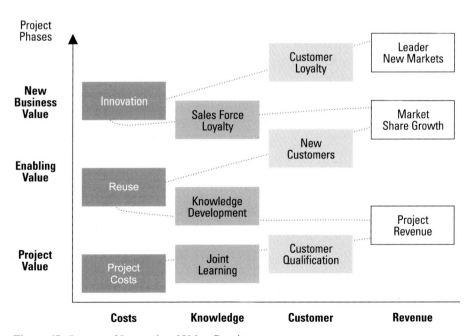

Figure 65 Process of International Value Creation

Enabling value

During the initial step in the value-creation process, the common project will lead to substantial costs, while there is no guaranteed Return on Investments. This is especially true of the visible costs of projects that involve international task force members. These costs may exceed those of local projects. International experts, however, know how to save time and, therefore, project costs. Furthermore, during an international project, all participants gain additional knowledge while working together in a team with global experts. Such teams possess far more knowledge than merely that of their day-to-day business and this is the knowledge that a local team cannot normally access in isolation. As an international collaborative project progresses, its potential revenue increases significantly.

The question of whether or not a project should be a collaborative effort has two possible answers. On the one hand, involving the right experts at an early stage in the business-development, or solution-creation phase of a project, may help to define the correct and most complete solution by making the customer aware of additional options. On the other hand, the likelihood of securing a project increases with the availability of additional expertise, making the projected revenue more likely to materialize.

New business value

After completion of a project, the knowledge gained during the process may have been a multiplier, a knowledge builder, within the local company. This means that both local companies increase their value. In addition, the knowledge gained in one project could be reused in other projects. Better qualified employees and the availability of local projects could enable the company to offer a wider range of solutions. This, in turn, helps to secure new customers in new market areas, thus increasing the company's share in the new markets and, simultaneously, improving the company's competitive advantage in local markets.

An international environment, open to different opinions that stimulate new insights, may lead not only to increased customer satisfaction, but also to increased employee satisfaction. This is the ideal environment for fostering innovations which are essential for keeping up with the competition in volatile markets. Companies may also see an increase in customer loyalty which, in turn, helps to raise barriers against new market entries and strengthens market leadership.

It became clear to the knowledge managers at ICN that the flow of knowledge would create additional revenue for both companies. However, this value had to be made explicit so that single units within the companies would become motivated to collaborate. But, how would this value be measured?

The companies' diverse motivational factors, influenced by the business environment, required a differentiated model – one which would present value creation in a better way than in the common business model. Put differently, a model that was sensitive to the diversity represented in the various stakeholders had to be found. While the relevant costs, for instance travel expenses, were a rather easy thing to quantify, the value of the (intangible) knowledge flow, opportunity costs or the future project value were extremely difficult to calculate. To assist with this problem, ICN had a closer look at traditional transfer-price models, before developing their own alternative, called the Knowledge Network Model.

The limits of traditional transfer-price models

Traditional transfer-price models served as a starting point for attempting an adequate visualization of global value creation. Transfer prices are, in general, calculated for resources transferred internally between single business units, or the various divisions of one firm. Their application aims to utilize the market mechanism within firm boundaries and, therefore, to accelerate a distribution of economic resources. Transfer prices serve as a means of linking company results to the their creators within the company.

As quasi-virtual prices, they can be based on such diverse figures as actual costs (personnel or transport costs), for instance, or services actually provided for a project, or opportunity costs. In the case of global collaboration, the latter would include the costs that arise from not having an expert available to create local revenue. In contrast to purely cost-based transfer prices, the price may also reflect the service, or working value, created by a person transferred to a certain project. A country, or business unit may also consider lower transfer rates based on what the anticipated added value will be, rather than the actual created costs and customer value. The transfer price is then based on the overall value that the knowledge-providing country eventually expects from the collaboration.

Siemens ICN examined the following models closely in order to make global value creation transparent:

1. Firstly, *market-based transfer prices* are calculated exclusively for prices of certain products or services that are relevant to the actual market, as in the case of independently operating profit centers. While the function of the market mechanism is fully transferred to the firm, market-based transfer prices only act as an optimizing mechanism, if an ideal market exists for the transferred resource, and both parties have direct access to this market. If this is not the case, incomplete information, or poor market conditions make the balancing of the market mechanism ineffective. This is true only for the actual costs of the collaboration, as this model cannot capture shared knowledge, or more abstract project values.

2. Secondly, *absorption cost-based transfer prices* are based on both the variable and the fixed costs of internally transferred resources. Although the knowledge-providing country is reimbursed the average total costs, the application of this model may lead to faulty decision-making as other costs (besides the decision-relevant variable costs) are also taken into account. This model is, therefore, most suitable for visualizing the transfer of resources that cannot be validated by a market and that are centrally employed and planned. (However, this presupposition cannot be assumed in the case of globally collaborating companies, as each local company plans and distributes its resource allocation individually.)

3. The application of *marginal cost-based transfer prices,* the third model, leads to debiting the knowledge-acquiring unit for variable costs. The fixed costs, therefore, remain with the expert-providing country. From an overall perspective, this model guarantees a maximum profit, but its application is restricted to the application of linear cost curves. The knowledge-providing country may have to deal with net losses on a permanent basis, on account of the lack of overall coverage of its fixed costs. Consequently, this approach cannot be used effectively to motivate both parties to collaborate.

4. The fourth model constitutes *opportunity cost-based transfer prices.* In the context of this approach, prices represent the knowledge-providing country's missed profits. An expert is not present and therefore not able to create local revenue during the collaboration process. Although this model calculates fairly accurate prices, it tends to assign profits to the knowledge-receiving country and does not provide strong motivation for the knowledge-providing country.

All the discussed transfer-price models above, are options that Siemens could have used for its international knowledge sharing and collaboration. However, not one of the pricing models considered was capable, on its own, of adequately reflecting the initiated current and future value-creation process. None of the models were regarded as an optimal stimulus for the cooperation of local companies. Siemens then considered combining some of the above pricing models which, in turn, led to the development of Siemens' own approach: a Knowledge Network Model. This new model will be considered in the following section.

Moving beyond traditional transfer-price models: Bonus on Top

The inadequacy of traditional transfer-price models led Siemens ICN to create its own approach, which is called Bonus-on-Top. Not only does this approach accommodate all their requirements, but it also makes the mutual and joint value-creation process transparent, thereby stimulating all parties to participate in international knowledge-sharing projects. The aim is for ICN to gain considerable benefits, such as knowledge creation, or the exchange of expertise and skills, from the collaboration process.

The Idea behind Bonus-on-Top

This collaborative process was based on several assumptions. These were:

- Each local company had a different area of expertise and, therefore, a reputation as a Center of Competence which could be helpful to other countries during the creation of customized telecommunications solutions.
- Local differences, like price levels or currencies, could prove to be inhibiting factors in the process of mutual knowledge exchange.

- The exchange would not necessarily happen in a bilateral way, but rather in a multi-lateral or network-like way, as some countries would be on the receiving end and others the providers of expertise.

In order to accommodate these assumptions and the diversity, ICN was given a tailor-made incentive system. *Premium-on-Top,* the first part of the scheme, had to be utilized to give global knowledge sharing an initial boost. The foundation for true knowledge networking had to be laid so that knowledge sharing and reuse became part of the daily job.

Bonus-on-Top, the second part of the scheme, was created to reward and instigate management backing of international knowledge networking, and was based on the local unit's impact on inter-company development. Thus, the goal of Bonus-on-Top is to reward the successful utilization of *ShareNet,* which generates international business revenue for ICN. This includes either revenue created in the company's own country, together with substantial knowledge from another country, or revenue created in another country, with the help of the local company's knowledge.

Implementing Bonus-on-Top:
Rewarding international knowledge networking

To launch the Bonus-on-Top as an incentive scheme for all ICN units within a local company, the management team of each unit was authorized to apply for the bonus, after receiving approval from the ICN board.

Bonus-on-Top comprises two complementary parts. The first part, the ICN Management Premium-on-Top, rewards a country's overall participation in global knowledge-sharing projects. In order to be considered for the bonus, the overall-achieved revenue through international collaboration has to total at least 5 percent (not exceeding 30 percent) of the local revenue. The management team is then awarded a bonus of approximately 10 percent of their salary, payable in the local currency (see Figure 66).

A central review board has been set up to assess whether a company has achieved its goal and qualifies for a Bonus-on-Top at the end of the fiscal year. As the assessment process requires some clarification, each country has to describe its project participation, and the resulting knowledge impact, using a standardized report.

The special Bonus Award for the top five international best-practice projects forms the second part of Bonus-on-Top. This second award not only rewards created value, but also acts as an added incentive for companies to share their knowledge, as it rewards creative and intelligent, general ideas about global knowledge sharing and reuse. To heighten the whole ShareNet community's awareness of stimulating projects, an application for the Best-in-class award has to be nominated by the community itself. A decision committee then selects the five most striking projects of the past fiscal year, based on best-practice quality. Unlike the ICN Management Premium-on-Top that is a reve-

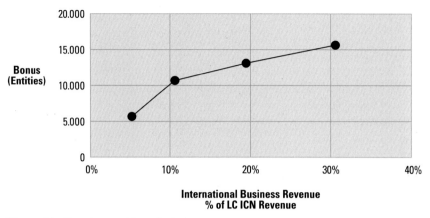

Figure 66   Premium-on-Top calculation

nue-based reward, these awards are of a special nature, like an executive management excursion.

You may be wondering what this has to do with Doug and Markus. Well, Bonus-on-Top was not only a direct financial incentive for the collaboration between ICN South Africa and ICN Germany (supervised respectively by Doug Williams and Markus Schmid), but it also reflected their common success and leveraged both parties' motivation. After completion of their common project, Doug reflected:

Receiving some management award naturally serves as an incentive to sharing our knowledge with our colleagues worldwide, but it is not the most important aspect. Getting direct recognition for how much our daily job is appreciated, is the most important thing. That's what counts and motivates us to carry on.

The Management Premium-on-Top award was not their only reward, however, as they were also nominated for the Best-in-class award. The necessity to record the process of their banking-solution project, in order to qualify for the award, also forced the ICN employees to reflect more deeply on their collaboration. This step leads to the externalizing and clarifying of many points in the process, exposing both the positive points, as well as the deficiencies, of a collaborative effort. In future projects these lessons are very helpful.

Perspectives

During his first day back in his South African office, Doug received two emails that were directly related to his collaboration with the Siemens headquarters. The first is from his German colleague Markus Schmid:

I hope you had an uneventful flight back home. Thanks again for your cooperation. I am already looking forward to our next common project! By the way, it looks as if we could modify our banking solution to fit a similar request that I have outlined briefly in the attachment. Your comments would be most welcome.

The second email was from Portugal:

Dear Mr. Williams,

I am urgently looking for some banking-reference solutions for a fast growing, innovative investment bank with a Hicom 300. This is for a good customer of ICN in Portugal, in the voice area of our business. We would very much like to get a foothold in the investment banking area, with a view to expanding our business. Our account team is looking for examples of Siemens expertise in this field. Your help would be much appreciated, Miguel Oliveira.

This letter bears testimony that fostering a culture and environment where knowledge sharing and reuse are all part of a day's work, is the basis for true value creation. The willingness to share knowledge reduces costs and helps to secure new business opportunities, to the benefit of the global ICN organisation. Management support of individual local companies is a key factor in facilitating this. The Bonus-on-Top scheme has proved a useful tool for promoting management backing of international knowledge sharing, as it rewards prolific knowledge sharers.

ICN knowledge managers are curious as to where global collaboration and knowledge sharing could lead. Although Bonus-on-Top was successfully integrated during its first year of application, the project is still in its infancy. The future scenario, however, definitely seems promising as ICN employees perceive Bonus-on-Top as a highly stimulating instrument. They not only like collaborating with their colleagues worldwide, but also appreciate the swift feedback and acknowledgement of their ideas and commitment.

Their enthusiasm is reflected in Siemens' success as one of the leading knowledge-sharing companies in the world. Besides a hundred other knowledge management initiatives, Bonus-on-Top yielded 250 million DM in additional revenue during its first year – all initiated by international collaboration.

Bonus-on-Top differs uniquely from knowledge management projects of earlier years in that it was developed at a grassroots level. It is but one example of how Knowledge Management is an operational answer to a fast-changing environment and how it leads to the emergence of a new direction on a strategic level which affects the entire company.

Add-on: Making Bonus-on-Top an integral part of employee remuneration systems

Bonus-on-Top was only the first step in establishing knowledge sharing at top management level. As the case study illustrates, it was meant to associate real value with the exchange of knowledge and experts. Bonus-on-Top, as a means of give knowledge sharing on a global level further impetus, only had an up-side without the down-side potential. Thus the regional managers received additional bonus payments, if they met a set goal, but if the goal wasn't achieved, the remaining incentives couldn't be accessed.

Indeed, the results of Bonus-on-Top were exceptional. During the fiscal year 1999/2000 revenue of 131 million Euro was reported as a result of projects obtained through international knowledge exchange.

This paved the way to the introduction of a strategic goal, namely international revenue through knowledge exchange, as an integral part of the annual wage agreements. This goal could influence the annual incentive both positively or negatively. Local company managers were usually rewarded for their Economic Value Added (EVA) contribution and other financial goals that they may have reached. International revenue is at present leveraging the financially-based incentive by increasing or reducing the earned incentive.

As a significant step on the long way to a knowledge-based enterprise the strategic goal of generating international revenue through knowledge exchange became an integral part of the headquarter's strategic controlling system. The strategic controlling system, namely the ICN Balanced Scorecard, was installed by the ICN top management as a holistic performance measurement at ICN. As part of the strategic controlling process, top management every quarter reviews the development of international knowledge sharing and reacts to red signals in the Balanced Scorecard with appropriate measures. For example, if a country with a high knowledge potential is not reporting high figures in knowledge sharing, special campaigns are launched within this country.

Staffing marketplace: PeopleShareNet

However, control is only one side of the coin and this cannot achieve much without supporting structures. International knowledge sharing had, in the first place, only been made possible by ShareNet, the international knowledge sharing platform. The Litmus project had shown that a virtual platform can only substitute our employees' face-to-face knowledge exchange to a certain degree. When it comes to mid-term projects, experts have to physically travel and work on a project full time.

As described in the business model paragraph of this case, the supporting environment is a key enabler of international knowledge exchange. Thus, ICN BTP decided to develop PeopleShareNet, a project aimed to facilitate people exchange on an international basis, thus enabling our employees to meet unencumbered by constraints of time and space.

As part of this initiative, a platform had to be created to match projects' needs and the availability of expert time and skills. A platform alone isn't sufficient for the seamless exchange of experts. Dealing with people means dealing with different motivations, a high degree of legal requirements and local working structures. Thus, the PeopleShare Net seeks to set up incentives for both the participating experts, as well as their managers who will be losing one of their best men to projects in another country. This will be done on a project-based incentive scheme, which allows all parties to participate in the project revenue.

The bottom line

The strategic goal of achieving higher revenues through international knowledge exchange has proven to be incredibly successful. Experience has shown that management buy-in and control by top management are key enablers of strategic changes at top management level. The Balanced Scorecard is an excellent option for a closed loop controlling system.

Metrics alone aren't worth much if the supporting environment isn't available to take appropriate action when goals aren't met. Dealing with people is a delicate matter and motivators can easily become demotivators if incorrectly set up. The win-win situation must be perceived as such by all involved parties.

Key propositions

1. **Customers and employees are first and foremost people and need to be treated as such.** Throughout the conception, design, development and implementation of this system, it was clear that systems and technology remain tools. They have to be implemented and participated in by people. The success of this project is in large part due to the consideration of what people really need – both as customers and employees. An organization neglects the human element at its own peril. This takes on special significance when there is widespread perception of the world and global organization in particular, as being impersonal and remote.

2. **Ongoing management support and feedback are essential throughout all the phases of implementation, as well as for ensuring ongoing success of a knowledge-sharing system.** Both explicit and implicit support from the top down, is crucial for the thorough and successful permeation of a new approach to sharing knowledge. The unspoken culture within an organization is a powerful tool for bringing about change, but the change of paradigm must be evident as emanating from the top.

3. **Increased contact networks, established for the purpose of exchanging knowledge, have the potential for more than just shared knowledge.** By facilitating personal contact beyond the limits imposed by natural geographic and service boundaries, an organization can potentially tap previously hidden human resources for use in shared projects. In addition, employees have an opportunity of promoting their own knowledge and skills without the restrictions of the traditional channels. They can literally "tell the world" of their capabilities.

4. **Systems like ShareNet introduce a new kind of meaning to the saying that "knowledge is power."** By sharing knowledge and expertise on a global scale, a single solution could yield many times its worth in a single application setting. The enormous potential for doing more with less, in terms of resources, remains to be fully realized as the scheme unfolds.

---

5. **The bonus system and stringent evaluation of new contributions are in keeping with modern scientific practices.** These include acknowledging the value of intellectual property (not always previously recognized), and of submitting work for evaluation by peers. By applying practices currently utilized in scientific practice, ShareNet contributes to the recognition of Knowledge Management as a science

Discussion Questions

1. Discuss the limits of traditional transfer-price models for a knowledge-sharing system.

2. How can you account for the phenomenal success of a system that initially posed a threat to employees? Critically discuss the motivations put forward in this case study for employees to participate in such a system.

3. In your opinion, is it realistic to expect the success of such a system to continue, or do you foresee the need for modifications to the system in the near future? Suggest possible improvements.

4. What criteria would you suggest for evaluating contributions by employees to ShareNet? Motivate your answers. (For students with an interest in Human Resource Management).

5. Do you anticipate that a system like ShareNet could potentially compensate for deficiencies in certain corporate structures (unrealistic production schedules, impersonal management styles, etc.) and emerge as a tool for affirmation, or have other similar benefits for the internal relations of an organization?

6. One of the objectives of this system is to build knowledge sharing into the workday of all employees. Given that extra time is seen as a rare commodity, what steps/strategies could you suggest for reaching this objective?

Four steps from knowledge networking to an organizational change

Thomas Klingspor & Felix Klostermeier

Abstract

How could Knowledge Management help to transform an organization? Answering this crucial question became relevant when the German market demand for Siemens' information and telecommunication products changed dramatically for the worse, and its sales unit required a solution to this crisis required. This solution entailed a change from selling products to selling solutions. Top management decided to employ knowledge management to support this change. The newly founded Knowledge Networking department started creating an infrastructure of knowledge management tools, thereafter measuring their local usage as well as the frequency of individual networking. The Knowledge Networking department finally used the benchmark figures to initiate a local consultancy. During this last phase it became obvious that related themes had to be brought in to create a methodology of focused analysis for change and control phases, thus answering the initial question.

Introduction

In October 2000 we visited a small sales unit within Sales and Services Germany, or Vertrieb Deutschland (VD). Our mission was to help overcome the employees' barriers to networking and communication. To start with we presented data about the main barriers. The figures were accepted by the unit's top manager who readily accepted responsibility for the upcoming change project and who then consequently named an employee, called the agent, to supervise the project.

This was Knowledge Management as we had always dreamed of applying it and it included the following four steps:

1. We had implemented a knowledge management infrastructure.

2. We had found a way to measure its usage and support.

3. We changed from being a tool designer to knowledge consultants (using the figures from step two).

4. We satisfied our (internal) customers' wishes by offering consultancy in other human resources areas (from innovation to information).

But before we elaborate on the steps, let's answer two questions.

The first one is: What is Siemens ICN VD? Siemens consists of numerous business segments, each representing a technical topic. One is the Information and Communications segment that consists of Information and Communication Networks (ICN), Information and Communication Mobile (ICM), and Siemens Business Services (SBS). ICM provides mobile networks and communication devices. SBS contributes solutions and services for mobile and electronic business. ICN offers products, solutions, and services for the Next Generation Internet, for example solutions for the convergence of voice and data networks, broadband Internet access, data and IP networks, optical networks, customer care solutions and mobile business solutions.

Siemens ICN has global sales divisions, one of which is VD or Sales Germany whose sales representatives are in charge of customers whom they advise regarding a choice from the newest communication solutions, while their colleagues from the engineering and service units install, implement and maintain these products.

The other question is, of course: Why did Siemens ICN VD need Knowledge Management? During the past few years the market for information and communication products has undergone a dramatic change. Whereas the customer yesterday only asked for the right product, he now needs a "solution provider". Again we can ask: Why? The answer is that comparability arising from the large number of competitors and the constant development of products, mean that the customer needs help not only with his telecommunication problems and needs, but with the determining of a telecommunication vision and strategy. He needs someone who understands his business nearly as well as he does.

The management of Siemens ICN VD saw a way of accelerating this market change by linking the know-how of their sales employees. This was, on the one hand, intended to reconnect the regionally structured sales regions and, on the other hand, it was becoming obvious that the massive, constantly increasing product portfolio was based on an enormous solution and product knowledge that no single employee could possess, but a specialist network could. In order to reconnect the regions and simultaneously create a network, the Siemens ICN VD management decided to initiate Knowledge Management.

In 1997 the knowledge management concept was introduced at Siemens ICN VD. A project group, consisting of persons with different functions and hierarchical levels, took an international inventory of the topic "knowledge networking" and used their findings to plan the first basic concept.

The first step: building a knowledge networking infrastructure

This concept entailed three initiatives.

To provide the decentralized organizational structure and the regional diversification with the needed knowledge, those employees who were willing to share their expert knowledge were linked: the KN (knowledge networking) Yellow Pages were born. To date 30% of all employees have entered their expert knowledge profile into the database. Every year 21,000 searches are performed, of which 30% lead to a satisfactory partnership, as determined by a survey and a tool analysis. Multiply 7,000 "positive" leads with half an hour's saved time and you'll realize what this tool is worth.

Individual knowledge about competitors and markets was compiled, after which it was editorially processed and publicized via the intranet for use by the entire staff of the various sales units. "KN Competitor Monitoring", as this initiative is called, has been successful enough to be turned into a department of its own.

To motivate service employees to share their knowledge, an incentive system calling for tips and tricks was initiated. Through this system service employees received rewards for valuable input. KN "Service Knowledge" saved a ratio potential of six million Euro.

All these approaches were based on the assumption that knowledge should be combined and exchanged in personal networks. Schumpeter defined these two generic processes – combination and exchange – as the two major processes with which to generate innovations. This assumes that new knowledge is also being created in these personal networks. It is the reason why we at Siemens ICN VD deliberately do not speak of Knowledge Management, but of knowledge networking. The goal is to find, to set free, and to enlarge the potential of current and new knowledge stored in the networks.

The VD-focused approaches to knowledge networking are now supported by an international ICN intranet platform called ShareNet. ShareNet enables salespeople from all over the world to exchange project experiences in discussion forums. In addition they can place "urgent requests", since this system allows them to instantly receive feedback from colleagues worldwide. In the backup database, salespersons can fill in and search for information on products, customers and competitors involved in a project. Parts of ShareNet's capabilities are also made available by the project-tool "Projektboerse" which focuses on the German market.

All of the above approaches are supported by various methods of communication, training and the implementation of knowledge networking in management instruments.

The KN project team is supported by a constantly changing group of undergraduates, graduates, and Siemens ICN VD employees who join the team temporarily while maintaining their strong connection with current research and practical experience.

Was this all that was necessary to support the change?

We were not sure. Although the figures confirmed the validity of our approaches, we were not convinced that the existing IT tools could indeed meet the needs of the thousands of employees who form a heterogeneous group. We decided to measure the state of knowledge networking in all the subunits that form Siemens ICN VD.

But making knowledge networks visible was no small goal!

We found the key to success not in a one-does-it-all solution, but in investing all our time and energy in an iterative approach – going for the partial solution instead of the total solution. This was how the KN Indicator was born[*].

The second step: measuring local networking with the KN Indicator

The KN Indicator measures the state of knowledge networking for each of the Siemens ICN VD units, differentiating between the objective and subjective measurement of networking. The subjective side is measured by an online survey of Siemens ICN VD employees, the first of which was carried out in August 2000. An online survey has the benefit of allowing anonymity and instant automatic evaluation. The objective measurement is made up of a usage measure (20%) that indicates the actual usage of the KN initiatives, and a support measure (30%) that allows a deeper insight into the actual employee support for the KN initiatives. This support, for example, includes the knowledge entries in the KN Yellow Pages, whereas the usage, for example, includes the recorded number of searches in the Yellow Pages.

Since the measurement of organizational knowledge networking had not been done very often, there were no benchmarks available within the organization. To compensate for this, single regional sales units served as benchmarks in the first cycles of the measurement.

The power of this measurement instrument became obvious when it was integrated into the employee/knowledge perspective of the Balanced Scorecard (BSC). Simply explained the BSC is based on the idea that there is a causal connection between the four business perspectives, namely employees, processes, customers and finance. From a single given strategy, different goals and drivers will be derived for all these perspectives. To illustrate: a formerly product-oriented sales unit wants to evolve into a solution and service provider (strategic goal). As a precondition for reaching this goal it is necessary that the sales employees be linked, since only a team can fulfill the needs of a complex product range (derived goal). The Yellow Pages are chosen as an effective method to support this formation of networks. It was surmised that when the network

---

[*] The KN Indicator was developed by Christian Dachs, Christiane Holz, Alex Rombold (former members of the KN project team) and Annette Laessig from Siemens Qualification and Training.

was widely implemented, the fulfillment of the strategic goal would be positively influenced, since linked employees would be able to use existing processes better, and create new processes, which would then lead to their being able to satisfy the customer's needs faster and better, whereupon they could demand more for their performance.

The success achieved in fulfilling these goals is represented by colors. Complete fulfillment is represented by a green light, and failure by a red light, everything in between is yellow. A red light accordingly represents an urgent need for action.

To react to yellow and red status indications in the KN Indicator, the knowledge networking team created a consulting process called KN Enabling. The task of this process is to support the local sales units with the implementation of actions to improve the local state of knowledge networking.

The third step: local consultancy to improve regional networking

The basic idea

Basically the KN Enabling process supports the management of a sales unit to reach the KN-relevant goals that, in turn, results in a green Indicator for KN. One can assume that a red or yellow KN BSC status is a result of barriers that obstruct the acceptance, usage and positive effects of knowledge networking. The KN Enabling process goes from identifying those barriers to deriving solutions to overcome them.

Our role in this process is consultancy, and this means a paradigm shift from our previous situation when we were thought to be fully responsible for knowledge networking at VD. It is no longer we who are responsible for the outcome, but the KN agent, mentioned at the beginning of our case, who is assigned by the unit's management to put the solutions into practice. This KN agent needs both resources and competence to run the KN process successfully. He receives these resources from his management, who should be interested in a successful process which will influence the important BSC positively. If the KN agent doesn't achieve what is expected of him... well, the onus is on the management. It is our job to link all former and current KN agents and supply them with state-of-the-art Knowledge Management. This helps in two ways: On the one hand the KN agents experience the power of networks themselves and, on the other hand, they receive advice from peer colleagues from whom they accept advice more readily.

In the end an enabled network will have been established that can apply Knowledge Management locally without our help. This is the goal of KN Enabling.

The KN Enabling process

The KN-enabling process starts and ends with the customer (local sales unit) which requires the KN-enabling support. The process is structured in five phases:

In the *initiative phase* the urgent need for action in terms of knowledge exchange and networking is identified through the KN Indicator. In the *analysis phase* the situation in the sales unit is investigated in greater depth. The information gained from the Indicator is used in this phase to identify the focus of the analysis. The results of this phase are the identified and interpreted local barriers to knowledge-flows. In the *concept phase* the instruments for overcoming these barriers are derived. They are put together in one conclusive concept which is presented to the unit management in the subsequent *decision phase*. After the approval of the senior manager, and eventual modification, the actions are put into effect in the *implementation phase*.

The KN Enabling object

The object of the KN Enabling process is the individual sales unit and its knowledge-flows. To specify the targets of KN Enabling more precisely, the unit is divided into clusters that are relevant to the knowledge-flow. These clusters are derived from the organizational structure. Examples of these clusters are listed in Figure 67.

The KN Enabling methods

No two sales units are alike, therefore the process has to be supported by a wide range of analysis methods and improvement actions. This is the only way to provide adequate action. Since the acquisition, processing, storage and sharing of this large range of methods and actions require a huge organizational effort, a "toolbox" was developed. This "toolbox" is available in database form and includes all methods of analysis, as well as all actions for improvement. The methods and actions are acquired by the KN team. References to these methods and actions from within the organization as well as from the outside, are investigated and documented in detail. The documentation provided is so user-friendly that even the most inexperienced KN agent finds working with these methods and actions does not require an intensive introduction or consultation. The methods and actions are then stored in the database with a tag leading to clusters

individual	Due to the constant increase in information, the employee is overtaxed, resulting in a lack of efficiency in fulfilling everyday work.
within teams	A lack of information sharing within the team results in ineffective cooperation.
across teams	Insufficient communication between departments leads to discrepancies and misunderstanding.
manager	Non-appreciation by the department head blocks knowledge transfer.
infrastructure	Inadequate infrastructure or wrong usage of tools may mean an unnecessary friction in the passing on of information.

Figure 67 Examples of the Enabling cluster

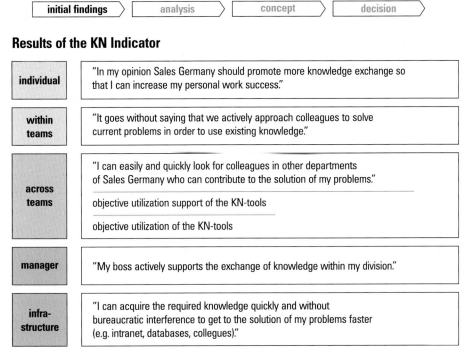

| initial findings ⟩ | analysis ⟩ | concept ⟩ | decision ⟩ |

Results of the KN Indicator

individual	"In my opinion Sales Germany should promote more knowledge exchange so that I can increase my personal work success."
within teams	"It goes without saying that we actively approach colleagues to solve current problems in order to use existing knowledge."
across teams	"I can easily and quickly look for colleagues in other departments of Sales Germany who can contribute to the solution of my problems." objective utilization support of the KN-tools objective utilization of the KN-tools
manager	"My boss actively supports the exchange of knowledge within my division."
infra-structure	"I can acquire the required knowledge quickly and without bureaucratic interference to get to the solution of my problems faster (e.g. intranet, databases, collegues)."

Figure 68   KN Indicator results (examples)

matching their aims. This allows the database to offer a selection of methods and actions for a specified cluster.

The KN Enabling pilot

In October 2000 our above-mentioned Siemens ICN VD sales unit showed the KN Indicator status presented in Figure 68).

The clustering of results made it clear that within this sales unit there were barriers to knowledge-flows between different departments, barriers in the support of the departmental head, as well as barriers in the infrastructure.

The analysis

In response to this indicator status, an analysis was carried out within the sales unit to gain a better insight into the situation. This analysis was supported by the thesis of a student employed by the unit, which included a questionnaire. The questionnaire was adapted to the identified clusters of barriers and reached 80% of all employees of the sales unit. The result of the survey is shown in Figure 69.

The survey result strengthened the findings we gained from the Indicator and justified a focus on the clusters of flows across teams, the manager and the infrastructure.

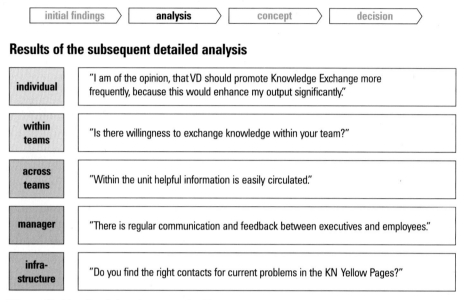

initial findings > **analysis** > concept > decision >

Results of the subsequent detailed analysis

individual	"I am of the opinion, that VD should promote Knowledge Exchange more frequently, because this would enhance my output significantly."
within teams	"Is there willingness to exchange knowledge within your team?"
across teams	"Within the unit helpful information is easily circulated."
manager	"There is regular communication and feedback between executives and employees."
infra-structure	"Do you find the right contacts for current problems in the KN Yellow Pages?"

Figure 69 Results of the subsequent detailed analysis (examples)

Arranged by clusters, these results lead to the assumptions listed in Figure 70.

The concept

During several meetings of the KN-enabling team with the KN agent of the sales unit, the results of the analysis were interpreted, whereafter solutions to overcome the identified barriers were derived. The actual actions leading towards the solution, were based on methods stored in the "toolbox" database, or were contributed by the employees. In the process of deriving these actions, the local conditions as well as the feasibility were taken into account. During this phase the KN agent was the main judge of those factors.

As an example of the practice applied, Figure 71 illustrates the methods, applied by the cluster "across teams", which have now been set into action.

individual	The green indicator status is supported by the analysis results.
within teams	The indicator, supported by the analysis, shows that the communication within the team works well.
across teams	While the indicator shows a yellow signal, the analysis leads to the conclusion that action has to be taken to improve the flows between the departments.
manager	Both the indicator and the analysis show a yellow signal and require action.
infrastructure	Analogous to the "manager" cluster, action has to be taken to improve the setup

Figure 70 Resulting assumptions per cluster

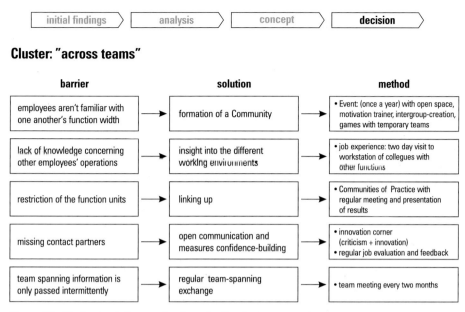

Figure 71 Derived solutions and actions for the cluster: across teams

The fourth step: transforming through the networking of management tools

With regard to the content of the Enabling object, the limitation put on knowledge-flows hindered adequate consultancy proceeding. It became obvious that the inclusion of existing management instruments in the KN-enabling concept was inevitable. Implementing instruments for the promotion of innovation without including, or at least considering, existing incentive programs, may serve as an example.

The idea of opening our analysis to other management tools came from the pilot unit itself when it requested the inclusion of the employee survey in the analysis phase of the KN Enabling process. Part of this survey included questions on the quality of communication flows. We gratefully adopted this suggestion and extended the focus of our analysis, as well as our cluster structure. The integration of information flows was the obvious following step. But where was this to end? The question arose regarding what system one would apply to include or exclude further areas, for example competence management.

We found a possible solution in the Intellectual Capital (IC) concept. According to the IC approach, the value of a company can be divided into the financial capital and the intellectual capital parts (Figure 72). The IC itself can then be divided into a relational part (as partly to be found in the BSC's customer perspective), in an organizational part (as partly to be found in the BSC's process perspective) and in a human capital part (as partly to be found in the of the BSC's employee/knowledge perspective). For us the

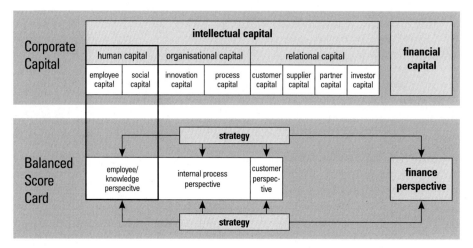

Figure 72 Corporate Performance Assets and their counterparts in the Balanced Scorecard

advantage of the IC concept was that the model provided a much more detailed look at the organization than the BSC perspectives did.

The most relevant part of IC for us, namely the human capital, can be further divided into employee capital and social capital. Employee capital represents the value of the individual (e.g. competence) and social capital represents the value of the network of employees (e.g. the network's potential to generate knowledge). These two company assets gave us the answer to our question about which areas to integrate into our analysis-consulting-controlling concept.

Employee capital consists of the competence, the readiness and the learning capability or flexibility of the individual employee. Social capital consists of a structural (structure of the network), a relational (assets inside the network) and a cognitive (accepted values and norms inside the network) dimension. These form the framework of our future approaches – approaches on which we want to expend effort, while improving knowledge networking.

Total quality management, competence management, innovation management, information management and Knowledge Management have to combine forces in order to reach their individual and common goals. Since all these fields have one goal in common: defined by the strategy, and controlled in the BSC, they have to support the organization in reaching its strategic goals. Working hand in hand will make this easier. The merger of the different processes is shown in Figure 73.

These days our activities are therefore concentrated on including the different fields of action into three parts of the KN-enabling process: analysis, consulting, and controlling. We have started to make the different measurement tools such as the employee survey, or atmosphere measure more compatible in a currently running pilot project, and have included them in our analysis. Our goal is not to create a one-does-it-all ques-

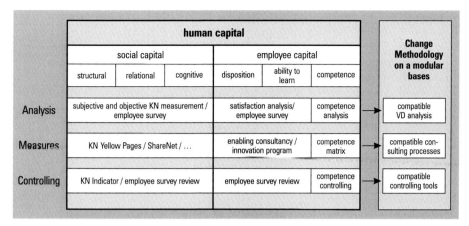

Figure 73 Transformation at VD – powered by focusing individual management approaches

tionnaire, but rather to make the different measurements more compatible in order to interpret them more precisely.

Similarly we are currently trying to get the different departments representing various management instruments into one boat. Innovation management, as well as total quality management and the controlling department immediately agreed to join forces.

The whole methodology of change therefore reads as follows: deriving the particular targets from the strategy, comparing those targets to the current regionally measured status with the help of a focused analysis and then overcoming the differences between the two measurements through an overall consultation, followed by a central and committed controlling of the methods adopted.

The running pilots show a very positive picture of how we have put these thoughts into action. This is not only a solution for the present Siemens ICN VD, but can be used to react to future market changes. It therefore seems that our goal has been reached and our journey is ended – for the time being.

Key Propositions:

1. The unit's top manager must be convinced of the approach and must have understood that he, not us – the consultants – is responsible for the success. Then he will automatically choose an agent who is dedicated enough to work even after we have left.

2. Surveys are fine to ascertain what is going wrong (within your questioned areas, that is), but they do not reach deep enough to discover why something goes wrong. We therefore recently enlarged our analysis toolbox with methods such as "social network analysis / critical incident" or "storytelling."

3. Only what is measurable is manageable. But the method of measuring must be chosen by the unit's manager. If he trusts in the Balanced Scorecard, fine. If he thinks EFQM (European Foundation of Quality Management) is better, why not?

4. An Indicator must be derived from the strategy and has to apply items for employees.

5. Focusing on the power of different departments mustn't mean having all of them doing the same, but making their efforts compatible. If this is understood, you no longer have to convince (or press) people to do something completely new (which they never will). Just ask them for help to open their methods in your direction – most of them will do this immediately.

6. Only a maximum of fifty to a hundred people can be helped to network daily. Knowledge networking for thousands at a time will not work.

Discussion Questions

1. What were the drivers for Siemens ICN VD to consider actions in Knowledge Management? Transfer these problems to other cases you know or have heard of! Can you identify the activities of the KN-initiatives at Siemens ICN VD in common concepts of KM such as those proposed by Probst et. al., or others?

2. What does Enabling mean in the context of this case and what significance does it have for you in a wider sense, e.g. in the context of organizational development?

3. Where can the "capital" be found within the human part of an organization? How does networking help to support and increase this capital? Discuss the relationship between these immaterial assets and the concept of shareholder and stakeholder value.

4. Discuss additional fields of management that you can you think of, which could join forces in order to increase their overall performance. Do you regard these collaborations as important?

5. Identify and compare the strengths and weaknesses of the approach presented in this case. Discuss possible improvements and barriers.

Knowledge Management for the e-business transformation

Albert Goller, Birgit Kleiber & Stefan Schoen

Abstract

The e-business knowledge management case illustrates the challenges that will accompany new business in an increasingly dynamic and global company environment such as Siemens and explores some of the fundamental directions that empower the e-community. This case study offers a short introduction to and an overview of the scope and elements of the e-business transformation within Siemens. Knowledge Management, which is one objective within this e-business program, is the topic and focus of this case. It considers the motivation and objectives of the e-community as well as the applied processes and methods in e-business projects. It likewise discusses relevant e-business Knowledge Communities throughout the world and Sharenet, the workspace for e-excellence. Aspects of learning and the motivation of employees are described as well as the effects of cultural change. Completing the case are some examples of results achieved with Knowledge Management for the described e-business transformation. This deliberately "open ended" case study closes with a reflection on the key propositions that arose from this project and areas for future improvement. This case study also illustrates the challenges and objectives of knowledge management programs that are relevant for Siemens.

Introduction

The Center of E-Excellence's primary objective is to quickly transform Siemens into a leader in e-business. The Center is responsible for planning and coordinating all e-business activities within the company and monitoring their success. An interdisciplinary team is developing the methods, guidelines and services required to meet this target. Other objectives include expanding the Siemens' knowledge base by promoting learning, knowledge transfer and best practice sharing, and by building an e-community within the company.

The e-business challenge for Siemens

We will become a leader in e-business by building a global value network of success. This success will have its source in the healthy flow of digital information and shared knowledge between our customers, employees, partners and suppliers along the length of the value chain of our worldwide business activities to wherever and whenever it is needed.

Siemens' Center of E-Excellence is a corporate-driven initiative under the leadership of the E-Business Council. The e-business strategy implemented by the Center of E-Excellence is comprised of 3 parts:

- Transform existing business.
- Create new business.
- Sell proven e-business solutions.

The E-Transformation Program was developed to facilitate the transformation of our current business processes to an e-business platform.

The mission of the Center of E-Excellence

In recognition of the central importance of e-business for Siemens, the Center of E-Excellence was founded in May 2000. It provides the impetus to transform existing business models, enables the provision of products and services, and creates new business through the Internet. The Center of E-Excellence provides corporate-level support for shared and existing e-business services and guidelines, facilitates the sharing of best e-practices as well as the mobilization of innovative e-business ideas. In short, the Center of E-Excellence enables the progression from traditional to electronic modes of operating within the company.

Scope & elements of the e-business transformation

The first objective is the development of an e-business culture within an e-community. Teamwork, speed, agility and customer focus – these are the essential elements of an e-community. To kick-start the development of our e-community, each Operating Group, Regional Unit and Corporate Office named an e-business representative fourteen days after the Center of E-Excellence had been assigned to assist with this goal. The e-community will leverage the benefits of e-business to develop more e-business for Siemens around the world. This means intensive communication across networks, working in virtual teams, developing globally consistent platforms, tools and modules, and exchanging knowledge and best practices.

The second objective is the development of an end-to-end, Internet-based value chain. This development is being pursued in cooperation with our partners and suppliers.

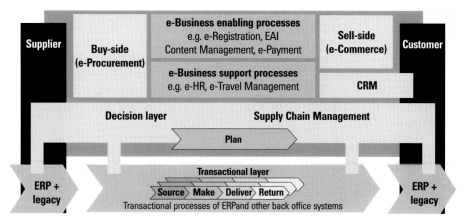

Figure 74 Scope and elements of e-business transformation

Whenever appropriate, new technologies are deployed to improve process quality, cut costs and simplify complicated transactions – all to the benefit of our customers.

The third objective is the need-driven exchange of knowledge and information. With e-business we are realizing the vision of Siemens as a knowledge-based company that uses the knowledge of our people around the world for the benefit of our customers. The business requirements along the supply chain – and they alone – dictate the content and structure of our future information flow. Knowledge should be exchanged freely among all our employees, both vertically and horizontally, unhindered by hierarchical structures, worldwide.

Our fourth objective is the defining of a set of binding IT standards for all Operating Groups and Regional Units. Three years ago Siemens also launched a special investment program of 250 million Euros to create the basis of a uniform IT hardware and software platform. Furthermore a high-performance network will ensure quick and simple communication between our employees, customers, suppliers and shareholders.

Within the next two years we will make an initial investment of about one billion Euros to facilitate our transformation into an e-company. The first stage alone is expected to result in enormous savings. A quarter of the total Siemens sales will eventually be transacted via the Internet – 50% and more of our consumer products.

The e-community knowledge management challenge

The successful realization of the Siemens e-business strategy, i.e. the transformation of the business, requires much knowledge which, at this moment, is scarce throughout the business sector worldwide. This knowledge includes, for example, know-how on new Internet technologies, process models, methods of business transformation etc. The rapid development and multiplication of the required know-how is a crucial success factor for the overall e-business initiative at Siemens.

The actual business environment is strongly characterized by globalization and distribution, cooperative ventures, speed and change, and knowledge intensity. For an efficient and effective knowledge sharing and creation, we need a professional network and the systematic collaboration of all employees actively involved in the e-business transformation from

- different organizational units at Siemens (groups, regions, corporate units),
- different business processes and projects, and
- different locations, time zones, cultures and languages.

The current situation implies a potential for huge economic benefits, such as savings on expenditure and resources, increased speed, or the reduction of risks through the prevention of:

- unintended duplication of work and repetition of errors,
- unexploited synergistic effects and effects of scale,
- failure to use existing knowledge within the organization,
- inefficient knowledge flows,
- a shortage of knowledge domain experts, and
- knowledge not being available in the right place and at the right time.

However, these challenges cannot be overcome by local knowledge management activities alone. Consequently the Center of E-Excellence decided to address precisely these problems with its project "e-community Knowledge Management" as the fourth step of the e-readiness program (see the section "E-business effects cultural change"). The project's objective is to accelerate the transformation of Siemens into a successful e-business company through systematic creation and reuse of its critical knowledge. This includes both experts' codified, documented (explicit) knowledge as well as their personal experiences (implicit knowledge).

The project has the following sub-goals:

- Creating transparency regarding all experts in critical e-business topics and facilitating access to their knowledge.
- Implementing the systematic exchange of e-business expertise between experts and projects.
- Creating transparency and access to all explicit knowledge assets (information).
- Implementing systematic creation and reuse of explicit knowledge assets (information).
- Supporting the standardization and implementation of best practices and generic processes.

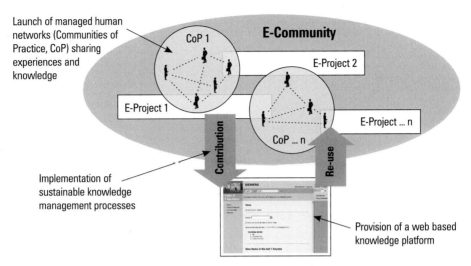

Launch of managed human networks (Communities of Practice, CoP) sharing experiences and knowledge

Implementation of sustainable knowledge management processes

Provision of a web based knowledge platform

Figure 75 Guided knowledge management approach

By developing the e-community, we want to improve the network between all employees (both experts and beginners) from relevant organizational units (Siemens Operating Groups, Regional Units, and Corporate Offices) who are actively involved in the Siemens e-business transformation. Members of the e-community are, for example, e-business managers, e-business project teams and Community members.

The approach applied to meet these objectives (Figure 75) forms the building blocks of the overall knowledge management concept, namely:

• the networking of experts in Knowledge Communities

• the imbedding of knowledge management processes into the e-business projects and

• learning for e-business, to expand the e-Community of possible matter end experts

The e-community will consist of a set of Knowledge Communities that are clustered around the Supply Chain Management, IT application and infrastructure, e-transformation and supporting Communities (e.g. e-communication).

The expected benefits

It is important to mention the expected benefits of this e-community knowledge management project. The reuse of knowledge is our key to success for innovation, time-to-market solutions and opening new business opportunities. It is an essential part doing business today and leads to:

• more efficient work processes as a result of the speed, quality and cost reduction that e-business projects provide,

• support for Siemens' process standardization,

- the gaining of a high acceptance since it involves the employees in Operating Groups, Regions and Corporate Units,

- an increase in Economic Value Added,

- saving on expensive training modules by offering internal knowledge transfer within the Knowledge Communities.

E-business effects cultural change

There will be new ways of communication and the information flow will be towards mutual knowledge sharing. The daily work will be focused more on interactive communication between virtual teams and Communities. Work structures are increasingly being directed toward exchanging knowledge on topics related to e-business. In future we will all work more transparently, however foreign this approach may initially appear to be. This altered method of working is nevertheless going to be our greatest challenge. Only teams with a communicative approach, capable of continuous development in an atmosphere of constant give-and-take, will be successful in the long term. This implies a culture of trust, appreciation and openness. Colleagues will learn to solve their problems together, avoid duplication of work and the learning curve of each e-community member will increase. Those who do not share their experiences, or share in the experiences of others, will eventually be unable to justify their presence.

Therefore it's important to realize that Knowledge Management for e-business related topics will dramatically change the cultural attitude of all who are involved in this challenging environment that stimulates the development of new skills, knowledge and attitudes related to e-business relevant topics. Cultural change is a continuous process that has to be constantly improved.

To promote and support such a culture change, the Center of E-Excellence – together with Corporate Personnel, Corporate Communications and Corporate Knowledge Management – has established the e-readiness project.

This project will ensure that the goal of employee e-readiness is gradually reached (Figure 76).

A technical prerequisite is that all Siemens employees worldwide will receive *Internet access* in 2001. Additionally, within the framework of e-readiness, a comprehensive *e-business training program* was developed to familiarize employees with e-business and to support their understanding of the Siemens e-business initiative. The *Employee Portal* is an especially important step on the way to becoming an e-business company. With their Portal, employees will increasingly be able to transact internal processes (e.g. self services such as HR-applications) via the Web. The Portal will be the central point of entry into world of e-business at Siemens and will support the information and knowledge flow within the company. E-community knowledge management and active knowledge sharing, which combined will support the Siemens e-business transformation, are the final stage of the e-readiness project.

The swift development and multiplication of the required know-how are crucial success factors for the overall e-business initiative at Siemens. E-community Knowledge

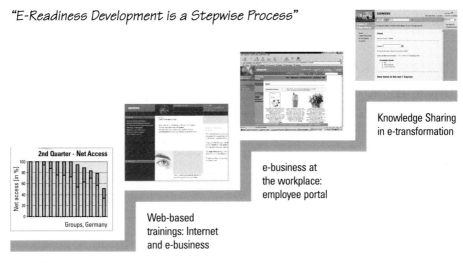

"E-Readiness Development is a Stepwise Process"

Knowledge Sharing
in e-transformation

e-business at
the workplace:
employee portal

Web-based
trainings: Internet
and e-business

Prerequisites:
Internet access

Figure 76 e-readiness

Management and actively driven knowledge sharing, as well as all of the employees involved in creating Siemens' e-business transformation, form one of the central elements of the change.

In the progression from "if Siemens only knew what Siemens knows" to "Siemens knows what Siemens knows", e-community Knowledge Management is an important milestone. To this end, and since e-business creates new knowledge every day, it is of the utmost importance that employees share their new experiences continuously. If they do so, they will be ensuring that in future knowledge is generated by projects with defined milestones. E-business expert circles (Knowledge Communities) will be started to assist such projects. They will do this by sharing their knowledge of important e-business themes and by publishing their know-how on the Siemens Knowledge Network.

Knowledge management solution modules

Processes and methods in e-business projects

The knowledge assets generated by the e-projects or Knowledge Communities should be available for reuse to all e-community members. The assets are, as is described further on, systematically picked up in the e-projects and Knowledge Communities using the Chestra methodology (see Figure 77), made available in a common knowledge database, reused and qualified.

Synergetic effects have to be fostered to utilize the ratio potential evolving from corporate knowledge and a corporate approach. E-business development is extremely fast paced and to ensure that the available methods and solutions are state-of-the-art, processes are required to create, improve, add to and maintain them.

The methodological approach

For this task, the Center of E-Excellence needed a best practice-based methodology. Best in class firms were identified and their IT methods were selected to develop the prioritized e-business domains (Marketplace Buy Side, Marketplace Sell Side, Supply Chain Management).

The various methods had to be aligned in a corporate approach: The ability to operate in defined roles with perfectly matched responsibilities and suitable skills while following outstanding guidelines, was analyzed as a key success factor. These goals could only be achieved within a consistent framework containing guiding principles, roles, processes and work-products. The framework with its corporate language would serve as a basis of all superordinate aspects, such as change management and management methods.

The Center of E-Excellence went looking for such a corporate method framework. Eventually Chestra, developed by Siemens Business Services (SBS), was selected as *the* in-house method framework.

Knowledge is refined in three loops:
1) Projects re-use and create knowledge.
2) Communities extract re-usable assets.
3) Method Architects distill methods from best practice approach.

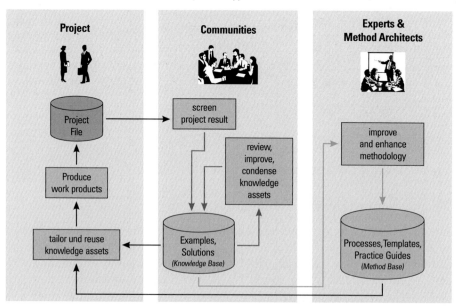

Figure 77 Chestra methodology process

Chestra characteristics

Chestra views a business problem from a holistic perspective: It contains a complete business change methodology that addresses all aspects of the business as an integrated whole. It therefore takes into account the challenges presented by business processes, organizations, locations, applications, technology and data.

The framework is naturally adaptable and tailorable. Chestra methodology components can be selected and customized for different levels of service contracts or engagements – from multi-enterprise to business area, and from large, complex solutions to relatively simple ones.

The framework keeps pace with emerging technology, business best practice, and changing customer needs. The Siemens Business Services Chestra Factory maintains the framework continuously: Processes, techniques, roles, and work products are frequently added to reflect new technical approaches, or types of business solutions.

Chestra also focuses on the human side of business change. Employees undergoing transitions often exhibit apprehension and resistance to change. Allowing change to "happen" rather than systematically managing it, often results in delays in implementing new and more effective processes and systems, unwanted attrition, and productivity short-falls. Organizational change seeks to identify those stakeholders who must urgently alter their behavior in some way to enable change, and then tries to understand their transition requirements, thus determining how best to support them in this process.

Knowledge Management and Chestra

Large scale knowledge sharing on a corporate level needs a common context: terms, concepts, processes, work product structures and values. The experts have to "speak a common language". It is especially true for the reuse of codified knowledge which requires a common basis. Since business, customers and markets change over time, Chestra too has to consistently adapt to these changes in order to keep its business value as well as to transfer know-how and the common language.

Experience and the sharing of knowledge are vital for the success of the transformation. To this end Chestra's framework offers a way to codify, store and reuse knowledge (Figure 77). This approach allows all users to search for and retrieve codified knowledge, thereby achieving the knowledge reuse scale required by a growing business.

Results from and experiences with Siemens' e-business transformation projects are the primary resources for the creation of knowledge assets. They form the basis for developing a methodology which is nothing more than documented and proven best practice.

Methods from the method database and examples from the knowledge database are used to create work products in projects. These are stored in project files. Some of them are selected as knowledge assets and published in the knowledge database. Some of the work products, combined with experts' knowledge are used to create or update methods for the next cycle.

In order to structure the collection of knowledge and expertise on a corporate level, the continuous knowledge evolution of the corporate methodology is based on the method framework and its common language. While practitioners and knowledge communities are building expertise and leveraging knowledge around a certain business topic, a collaborative environment will support them in leveraging best practices in their daily work. Knowledge asset creation processes are set up by the SBS Chestra Factory to capture expertise and to maintain this framework and the collaborative environment. A process was therefore created to align Chestra with these evolutionary changes. This process is based on a feedback loop that transforms individual experience practices into best practice.

Coaching as a project guidance

E-business transformation projects require an introduction and skills development of the methods, tools and guidelines relevant to the specific project at project-specific intervals and at project-specific intensity.

To this end the Center of E-Excellence offers a coaching program that provides methodology, knowledge and best practice sharing to support all transformation projects from their vision and strategy up to and including the required IT operations. The coaching program therefore encourages those involved in the project to make use of the corporate methodologies and knowledge assets.

The coaches are subject matter experts experienced in Siemens e-business transformation. They will support the project team by retrieving appropriate existing knowledge assets, help to create new knowledge assets and afterwards use the lessons learned for similar projects to prevent duplication of effort. They will furthermore organize for experts in training and integration of method to be present in the specific Knowledge Community.

Global Knowledge Communities on e-business subjects

The knowledge management approach, described previously, focuses on the formal process of generating and reusing knowledge assets in e-business projects. To address more people within Siemens who are involved in the e-business transformation, we decided to launch a second, perhaps less formal, approach. In the current Siemens business environment, we regard subject-focussed Knowledge Communities that collaborate across organizational boundaries, as an appropriate and promising approach to create and share knowledge on crucial e-business subject domains (see the Development of the Siemens Knowledge Community Support).

E-Community Knowledge Management became a subproject of the e-readiness program for which we defined the following goals: By developing the Siemens e-community we want to improve the network between all employees (both experts and beginners) of relevant organizational units, Siemens Operating Groups, Regional Units and

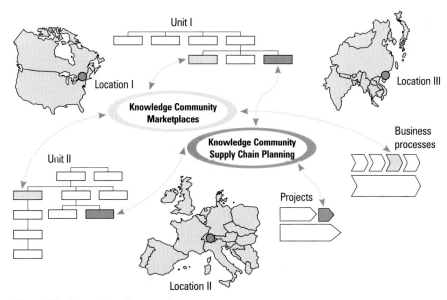

Figure 78 Knowledge Community

Corporate Offices, who are actively involved in the Siemens e-business transformation (Figure 78). They can make a valuable contribution towards optimizing the handling of the Siemens e-business knowledge and thus improve the business goal of a swift, efficient and successful e-business transformation.

With this in mind, we staffed a team consisting of people from different backgrounds, ranging from strategy and marketing to e-business and knowledge management. Within the first phase the team developed a common understanding and concept regarding a variety of aspects, for example:

• the motivation and objectives of the project;
• the introduction to roles, activities and outputs of Communities;
• the steps towards starting a single Community;
• the enabling activities of the e-community Knowledge Management and
• the project plan.

The general objective of each Community is to share and create knowledge and to obtain transparency about experts and information sources. In order to achieve these objectives, there are a variety of activities in which community members are involved. Different communication channels are used for the activities of a Community. A sensible mix of face-to-face and virtual communication is needed in order to exploit the respective advantages of personal contact and fast, straightforward communication between sites distributed worldwide. Right from the beginning there are important roles which have to be assigned, for example, the moderator, the sponsor, or the knowledge brokers.

We, as a team, conducted an analysis of the stakeholders and the business environment, created presentation material and saw to it that suitable management support was available. Simultaneously we kept contact with the e-readiness sub-project that, in parallel, was working on the Sharenet knowledge management workspace for e-excellence.

Early on in the e-community knowledge management project we started a pilot Community. The topic was "e-communicators" – the members of this Community work in communication departments throughout Siemens. Through this pilot Community we learned much that we could apply during the community roll-out phase during which we addressed topics like e-procurement, supply chain management, marketplaces, and e-strategy.

For the start-up and support of an individual community we developed an approach with three phases: 1) Preparation, 2) Ramp-up, 3) Sustaining & Improving.

In the preparation phase of each Community the initiating members of the Community have to be involved in the design of the community concept and the initial knowledge management activities. The most important aspects of this first phase are ensuring management support, finding and training a moderator, as well as the identification and early involvement of potential members. Moreover, the set-up of the IT platform and the collection of initial content are also important aspects.

In the second phase – the ramp-up phase – the knowledge management activities are usually focused on the creation of transparency regarding experts and projects, the sharing of valuable information (for example news, checklists, process models, as well as project plan and business cases and experiences) and the solution of daily problems (for example, the solving of urgent requests). These first activities form the basis of a motivated Community and should create activity in the Community regarding communication and the collecting of a critical mass of relevant and interesting information.

During the whole phase more members are invited and integrated into the Community – people from relevant project teams in the Siemens groups and regions, people with special roles like e-business managers, or Chief Information Officers and consultants from other internal units supporting the e-business transformation.

Depending on the members' opinions, we hold a face-to-face kick-off workshop as early as possible with no more than 35 persons (from normally more than 100 members in this phase).

In the third phase of each Community – the sustaining and improving phase – the goals are extended when compared to those of the second phase. In this phase the standardization of processes and methods, joint development of new innovative e-business solutions, and the reusing of whole e-business solutions from other projects are undertaken. Overall the goals become more ambitious as the Community advances towards the higher levels of the knowledge ladder, and the information turns into high quality knowledge. To achieve these goals, the community members interact additionally in small sub-groups on special sub-topics or subject-focused events.

In this third phase the group is again enlarged to include more of members. Although most of them are less active, this move still generates much benefit if these members sometimes reuse knowledge provided by others. In this phase further activities become useful, for example acquiring knowledge from external "think tanks", and making improvements to the Community based on the monitoring of activity and impact.

After the start of a Community you have to use the momentum of the start-up phase to breathe vitality into the community concept that has been developed. The objectives must be followed by sustained action and benefit to the business. The Community should become an important component in the day-to-day work processes of those involved. In addition, you must constantly adapt the objectives, activities and the enabler of your Community to the changing environment.

Some results of running Communities are described under the section Examples of the E-Communication Knowledge Community. The activities described therein are based on Sharenet, the e-business knowledge management platform.

Sharenet – workspace for e-business excellence

The Sharenet for e-business (Figure 79) is the knowledge management platform which is based on Internet technology and supports the e-business transformation. Its connectivity is the key to building knowledge networks. It features a unique connectivity between knowledge objects relating to a business process and is based on ideas originating from Siemens ICN (Information and Communication Networks) ShareNet, as well as Corporate Knowledge Management (IK CKM). Through Sharenet we want to network interacting communities that use, leverage and offer knowledge concerning the Siemens e-business transformation. The development of Sharenet for e-business excellence, started in August 2000. On October 10, 2000 the first version of the system was initialized.

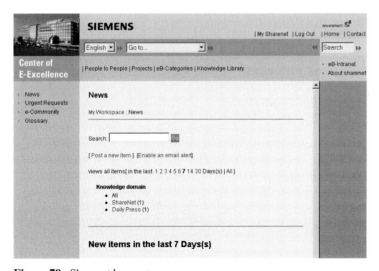

Figure 79 Sharenet homepage

Internet technology is the motor of an interactive communication between the different e-business projects and experts at various Siemens branches world-wide. Many collaborative functions help members to obtain to exchange experiences, best practices and last, but not least, lessons learned.

Project management methodologies such as Chestra, (see section "Processes and methods in e-business projects") are also integrated into Sharenet to support e-business projects. Additionally, Sharenet provides an international overview of most running and planned e-business projects within Siemens.

Since its initiation in October 2000 the database and different connecting services have been developed in swift, iterative improvement cycles. The next development steps will cover the ergonomics and the user interfaces.

Sharenet for e-business has different applications. These are accessible through the different views, via hot-link navigation and clickable maps, namely:

- "People-to-people" which includes discussion groups, community workspaces, overview about online-users and a chat room
- "Projects" which includes projects and methodologies for the e-transformation
- "eB-Categories" which gives and overview and provides access to knowledge objects and projects via mandatory and regular structure of e-business transformation
- "Knowledge Library" provides access to all knowledge objects which are filed in a structured manner
- "News" which includes news boards
- "Urgent Requests" which displays all urgent requests of various discussion boards
- "E-Community" provides an overview and possibility to search for members of the e-community
- "Glossary" contains abbreviations and technical terms often used in conjunction with e-business
- "My Sharenet" offers a function to change a password, personal user information, an overview of personal Sharenet shares, and objects and activities

The above mentioned points of entry are just different views of the knowledge assets (information). The knowledge assets are the bolts and "e-nuggets" which are the heart of the transformation engine. All information is stored in (knowledge) objects, this knowledge being codified knowledge which is thereafter called "information". A (knowledge) object is the smallest possible unit in which an information package can be stored.

These are then used by the target groups in order to guide them intuitively to the requested information. These views allow different ways of accessing the (knowledge) objects. It is also possible to sort the objects according their attributes. In order to have a common look and feel, some main views are defined and cannot be changed.

Knowledge Management for the e-community offers a perfect opportunity to transform Siemens into an e-driven company very quickly. The members of the Siemens e-community share their high-quality know-how and practical experiences with experts in other Communities in a global, barrier-free network that operates across Operating Groups and Regions. Access to the Internet-based knowledge sharing platform, 24 hours a day, 7 days a week will also be important.

What are the benefits offered by the Sharenet workspace for e-business excellence to people involved in e-transformation?

- An intensive, interactive exchange of high-quality information and know-how.
- Access to insights, best practices, experiences and lessons learned on e-business related topics.
- Access to most complete and relevant information sources for quick decisions.
- Discussions on roadmaps, strategies, and global trends.
- The opportunity to build with others a global, barrier-free social and competence network to enrich one another across operating groups and regions, and
- An opportunity to increase their personal market value.

Learning for e-business

Knowledge Management is mostly concerned with aspects such as the identifying, capturing, systemizing, bundling, and preparing of knowledge assets for reuse within a company. Learning management, on the other hand, is mostly concerned with the didactical formats of the content, the alignment of knowledge with business strategy, target group-specific offerings, and a swift, efficient, convenient, measurable and sustainable diffusion of knowledge subjects within a company.

While Knowledge Management for the most part addresses the development of the knowledge base, learning management's role is to present knowledge to a much broader Community, in other words, to expand the Community of possible subject matter and experts.

In this sense an effective learning management is dependent on a well established Knowledge Management for content with substance. Vice versa a valuable Knowledge Management uses a well established and seamless connected learning management to address a broad audience within a company swiftly and efficiently. Knowledge Management and learning management are two complementary disciplines that are continuously growing closer and support an innovative and agile enterprise.

The starting point for our learning activities in the e-readiness program was the simple question: "How can we efficiently mobilize and direct all the talent and power available at Siemens for e-business?" The answer was simple and obvious: By combining "Learning for e-business" and "Siemens' On-line Experience" by means of e-learning.

The ideas guiding the e-business program

- Learning is the mastery of self-improvement and must be synchronized with and integrated into the daily business environment. Self-driven learning must become part of the normal business life.

- Deep beliefs and assumptions can change as experiences changes, and when this occurs culture changes. Deep beliefs and assumptions are not always consistent with espoused values in organizations.
- Due to the construction of the e-business program, the coming demand for learning and change opportunities will vary in scope and scale over time. We therefore need to establish a concept that responds to the dynamics in demand.
- If there is a learning opportunity and change experience that can be effectively supported by e-solutions, it should be provided through the Net for availability, adaptability and efficiency's sake.

But... no learning without substance! Who is going to deliver the content for e-business at Siemens when the Knowledge Management of e-business hasn't as yet been fully established? Siemens Qualification and Training came up with the perfect idea. They established a content group that drafted the story-line, captured primary content from subject matter experts and Communities and transformed it into a more communicative and on-line teaching mode.

One of the results of this activity is a web-based training program *From Siemens to e-Siemens* that was developed by the e-readiness program team. It reflects the cornerstones and key success factors of the Siemens e-business program and presents our approach in a comprehensive, compact, convenient and efficient way to the company as whole – on-line via the web, of course.

Now it's up to each Siemens employee to step into the field of e-business and join Siemens' e-community.

Motivating employees for e-business transformation

Sharenet Shares

There are ways, however, to motivate people to exchange their knowledge. Within the Sharenet workspace for e-business excellence, a combination of intrinsic and extrinsic motivators are used. Carefully crafted workshops introduce users within the Knowledge Communities to the advantages of knowledge sharing. Working in Sharenet also offers a unique bonus system. By contributing knowledge, a user can accumulate "Sharenet shares", which can be exchanged for a variety of knowledge-related events. On the whole Sharenet is currently an enticing, living system with the potential of attracting many members.

Only when we have made up our minds that sharing knowledge is important, not only for efficiency's sake, but also to increase the essential humanization of the business and social environments in which we work, will we be prepared for the tasks confronting us.

Examples of achieved results

Sharenet facts:
Within one year Sharenet has already achieved respectable results.
- more than 3.400 users
- more 2.400 knowledge objects
- more than 420 e-business projects
- 214 teams
- 134 discussion boards

Example of reuse feedback and comments

More than 100 reuse feedbacks and a hundred comments have been added to knowledge objects or e-business projects in Sharenet. For example, a project from Siemens Russia was published in Sharenet where after I&C (Information and Communication) Norway was interested in obtaining more detailed information on this project, since a customer of theirs required something similar.

A second example illustrates a comment which has been added to the Knowledge Object "A&D Intranet based Call Management System":

["Hi T., I read with interest about your recent successful launch of the web-based call management system and I'm keen to find out more. Is this a self developed solution? What are the costs like? *I see a possibility of reusing this solution,* as we are searching for something along this line...where could I learn more about this setup and the features? Regards, W."]

Both examples shows, that colleagues from different organizational units are working on related projects and they can adopt each other's solution.

Example of the E-Communication Knowledge Community

The first of the Communities of Practices were started on the initiative of colleagues who, at the junction between the Center of E-Excellence and the Corporate Communication, had to deal with

- providing colleagues with information on e-business,
- further empowering colleagues and management to use the Internet and intranet,
- the change that was occurring regarding international cooperation within the company with the support of the newly developed tools, and
- improving the effectiveness of colleagues' cooperation with experts and partners on a daily basis.

This Knowledge Community was called E-Communication Knowledge Community and it worked on its first common topics at a Kick-off meeting in Beilngries, North of Munich, on 22 and 23 February 2001. The first work meeting was to take place in the vicinity of Zürich on 17-19 July 2001. Twenty four colleagues from 11 countries and 8 operating groups such as corporate departments, participated in this meeting. The following goals were defined for this Community:

- The joint development of tools with many applications, materials (content), media and criteria for the measurement of employees' e-readiness.
- The joint development of e-readiness communication activities, so that existing concepts of e-readiness can be controlled and applied.
- Setting up of a synchronized roll-out plan for the worldwide E-readiness Communication and an estimate of the expected costs and time requirements.

During the Kick-off meetings three work groups were formed for the themes:

- Internet access for employees.
- Examples of best practice and how we can communicate their effectiveness best.
- Communication tools and methods.

Not all of the participants, however, were willing to accept or sacrifice time for follow-up tasks. Although everyone present contributed to the Kick-off meeting, at the end it seemed as if no-one wanted to commit themselves to the completion of "homework" and yet, afterwards most did.

The following are one negative and two positive examples of occurrences from the time between the Kick-off meeting and the preparation for the first work meeting:

1. When the moderator first suggested a Community chat session and invited the members to communicate directly with one another in a smaller group, there was at first very little interest. What was this reason for this? The theme was interesting, therefore this could not have been the problem. It became apparent that a technical server error as well as the unfriendliness of the user interface were to blame for the failure of this first chat session.

2. When colleagues in South Africa asked material for an employee information action, i.e. simple, group-targeted texts in an approved form, they first of all telephonically requested these texts from Headquarters. When the moderator presented the South African request to the private discussion forum of the Community, the lines began to hum. It became apparent that at various places there was already much material available, or in the process of being printed. The texts that were the answer to their request, were available to these colleagues at the community workspace in Sharenet on a common basis.

3. When the first colleagues from various parts of the world requested pre-drafted parts for employee brochures, posters and organization equipment, the planned E-Readiness Toolbox was almost complete. After the various countries' requests had been discussed in the private discussion forum of the Community, there were two results: the first was a pre-release, but applicable, part of the Toolbox for the exclusive use of Community members and secondly a very fruitful, closed discussion about the completion or changes to the planned Toolbox.

Conclusion

Top management support is tremendously important. Only if the management supports and also lives Knowledge Management actively, it will work within the organization as a whole.

Starting the e-community we discovered that it is of the utmost importance that the knowledge management support team provides full assistance to all members of the e-community right from the start onwards.

All employees have to be informed about Knowledge Management and they have to know why sharing knowledge is so important for the e-business transformation. They should also be shown the benefits that this will have for themselves. Another important point is that there should be special training in the use of Sharenet. It should be made clear that there are different applications possible and not only one point of view on knowledge objects. It is necessary to explain the structure, where to find public information and knowledge objects. Information is only valuable if it can be found easily.

The members of the e-community also have to understand that they are responsible for updating their published knowledge objects or deleting them if they are no longer relevant.

Key propositions

1. Knowledge management initiatives have to be aligned with corporate goals. At Siemens these objectives involve the transformation to an e-driven company.

2. Top management involvement and commitment are tremendously important and a prerequisite for the successful implementation of global knowledge management initiatives. Management can promote knowledge sharing by repeatedly stressing its importance for the whole company.

3. To ensure the global reach, a professional network and the systematic collaboration of all employees who are actively involved in the e-business transformation and the new communities have to be supported in the starting phase.

4. The swift development and multiplication of the required know-how are a crucial success factor for the overall e-business initiative at Siemens. Therefore efficient and effective knowledge sharing and creation have to be practiced continuously to overcome structural and organizational barriers.

5. Internet technology is the motor of an interactive communication between the different e-business projects and experts at various Siemens branches world-wide. Therefore an Internet-based knowledge management tool like Sharenet, the workspace for e-business excellence, offers the possibility of sharing knowledge 24 hours a day, 7 days a week.

Discussion Questions

1. State reasons why Knowledge Management for e-business is important for companies like Siemens.

2. Describe the main requirements of a knowledge management solution in the given situation.

3. What are the main benefits of knowledge sharing related to e-business relevant topics within Siemens? Specify three points.

4. Discuss the followed e-business approach. What risks and barriers do you see endangering a successful and sustainable implementation of the process? What alternative approaches or modifications would you regard as useful?

5. What are the new challenges and opportunities the web poses in this process?

6. Discuss effects of the cultural change and how they will influence the daily work of the employees.

VII   Epilogue

Putting knowledge to work: Case-writing as a knowledge management and organizational learning tool[*]

Gilbert J. B. Probst

In search of effective knowledge management tools

The meaning of knowledge management to Siemens has been thoroughly demonstrated by the case studies discussed in this book. In fact, the common theme in all these contributions is the quest for answers on how to make knowledge work for a company in the new economy. It has been made clear that Siemens understood that its knowledge resources required great care, as well as approaches that sometimes differed radically from management of traditional organizational resources, to put knowledge to work and, ultimately, to profit from it.

All this is well known in theory, but my experience shows that few companies are eager to follow through the implications of these insights in practice. Typically, companies are reluctant to go beyond installing IT-based platforms for sharing knowledge, often ending up realizing that these platforms are not used to their full extent. These platforms are often simply superimposed on existing organizational structures to alleviate the dysfunctional effects these structures have on knowledge sharing. The problem is that these IT solutions often gradually develop into appendices to everyday practice, rather than becoming intertwined with practice.

The practical approach that inspired the present book makes it unique among many conceptual studies and theoretical reflections. It is in this spirit that I would like to end this book. My aim in these last pages is to introduce the reader to a practical tool and to comment on the method used, writing cases as a learning tool, a way of creating transparency and interpreting, creating, acquiring, developing, transferring and retaining knowledge. The building blocks of managing knowledge, as I have developed and researched with my colleagues in Siemens as well as many other companies as part of the Forum for Organizational Learning and Knowledge Management, are well illustrated in this case writing. Any company in any industry can use this tool, irrespective of the systems at hand. It is a knowledge creation and transfer tool not affected by any

---

[*] Parts of this contribution are based on a presentation at the 6th Annual EDINEB International Conference, June 1999, Norwegian School of Economics and Business Administration, Bergen, Norway by Gilbert J. B. Probst and Claudia Jonczyk

IT systems' integration; it is relatively cheap and, most importantly, puts knowledge to work effectively. This is not some pipe dream, but a tool that Siemens has put into practice throughout this book.

Why case writing?

Cutting-edge companies are increasingly using case writing as a convenient tool for putting their knowledge assets to work. These companies have found that in the normal course of their work, members of project teams acquire and share knowledge about managerial practice. When the project has been completed, however, the teams disband, individual knowledge is dispersed, and collective learning is often lost. There is little incentive to reflect on the accumulated experience or to document it for reuse in future.

Companies have tried to prevent the loss of this knowledge, and the associated costs of repeatedly re-inventing the wheel, by storing it in databases. Investment in IT solutions, however, does not always bring the expected rewards. There are many possible reasons for this lack of acceptance: the database may be too complicated; employees may not trust it; the idea may conflict with company culture or the prevalent management style. The consequences are all similar: knowledge is not transferred to other employees who might be able to benefit from it. Under these circumstances, the organizational knowledge base does not evolve; it does not renew itself. There is a distinct possibility of the organizational knowledge once again becoming an inert asset.

This is a serious problem, especially in the "Internet age", where incumbents' rich organizational knowledge constitutes perhaps their most important competitive advantage over start-up companies. The current compression of product and communication cycles, the blurring of industry boundaries, and the deconstruction of value-chain market models that are "blown to bits", as the researchers Evan and Wurster put it, increasingly jeopardize the existing value of the organizational knowledge base. Many companies are unaware of this, continuing as if everything were in order. Michael Beer of Harvard Business School even compares this process to the lethal effects of cholesterol, and talks about the "silent killers".

To cure the "silent killers", a process of organizational or collective learning is needed that will allow new knowledge to influence company activity, thereby sustaining the value of the accumulated knowledge. In the recent past, a number of approaches have been developed to address this problem, and greater emphasis has been put on the "human element at work", as opposed to IT solutions. It is in this line of thinking that the practical benefits of case writing become obvious.

The benefits of case writing

Case writing is mostly practiced in management training. As a teaching tool, cases are widely used in MBA programs around the world, since they enable students to learn from real-life situations that they, as future managers, will encounter. Working with cases gives students an opportunity to compare their own solutions to problems with the actual ones. Discussing and evaluating alternative possibilities help them to acquire a wider view of realistic management options. In this sense, case studies are vehicles for transferring company knowledge and experience.

What is more, due to the narrative style of case studies, they are open for discussion and reflection. In fact, this is precisely the rationale for using them for teaching purposes in the first place. This suggests that cases are sensitive to the different types of knowledge contained in a particular business problem. Not only is conceptual knowledge being recollected, but also the practical experience of putting this conceptual knowledge to work is conveyed. Through their ability to convey intricate problems and experiences, cases are uniquely suited to portray the tacit knowledge and experience acquired over time. Not only do tacit best practices and common experiences become explicit, but new light is also shed on past failures and past problems are disclosed. In this sense, case writing enables organizational learning – it forces the writing team to collect knowledge, to reflect on it, to make it explicit and allow transparency, to show its importance and applicability, to think about sharing, distributing and retaining it in form of a case, and the use of the case in-house. In parallel, this method can also contribute to intensifying cooperation and teambuilding, as participants have to think about who might be interested in what kind of knowledge, how to transfer and teach the relevant practices, and it might be important for "socializing" new organization members.

Given the benefits of case studies for teaching purposes, it is interesting to note that in industry, except for a few cutting-edge companies, little use has been made of cases as a method for management training. I am in no way suggesting that companies should organize seminars or workshops to study case histories taken from other organizations. My proposal is rather that companies should systematically write their own cases, i.e. jointly document the knowledge and experience they themselves have acquired over time. During the joint documentation process, knowledge that is implicit and closely linked to experience can be made explicit and put to work. The narrative style of case studies makes them infinitely more interesting and engaging than the ubiquitous bulleted presentations that pervade corporate life. Finally, and perhaps most importantly, case writing within a company has the benefit of truly fostering organizational learning. The collective experience of jointly recapping past experience allows for levels of retrospective sense-making hitherto untapped.

My research shows that case writing, as an organizational knowledge management tool, is associated with the following benefits:

- Case writing is a convenient tool for collecting, creating, transferring and retaining knowledge in the company.

- Due to the narrative style of case studies, case writing is sensitive not only to codified and explicit knowledge, but also to the rich experiences and tacit knowledge acquired in the past.
- The sheer process of collectively writing a case study about one's own experiences fosters learning on several levels of the organization in an upward spiral, starting with the individual, proceeding to the group, and then corporate levels.

I will elaborate on what I mean by this in some detail in the following pages, before actually turning to the key success factors that could make case writing work for your company.

A tool for collecting, creating, transferring and retaining knowledge

Case writing is a convenient tool for transferring knowledge for primarily two reasons.

Firstly, the collective process of case writing in itself fosters the retention of the knowledge transferred. Also critical to this process is the transferring of relevant knowledge and the continuous challenging of the value of the knowledge through the "devil's advocacy" role – played by the team in the process of joint writing.

Secondly, the product of case writing, i.e. the case study itself, through the richness and depth of its characteristic narrative style, is a much better vehicle for knowledge transfer than bulleted lists.

Case writing as a collective process

Why are cases a convenient method for transferring knowledge? As the value of transferred knowledge is solely dependent on being received and retained, a closely linked problem is how the retention rates of the receiver can be increased. The answer is found in the psychology of learning. Empirical research in this field of learning and memory indicates that retention rates are dramatically increased when participants play an active role in the learning process. Actively producing and recording information can result in a retention rate of approximately 90%, compared with only 20-30% for material that is simply heard, for example, in a lecture, or for material that is heard and visualized (in a lecture with transparencies or videotapes). These findings have promoted a growing recognition of the "action learning" approach in management, in which learners have to combine learning with action. The learning process starts with practical experience, which is then subjected to a process of collective reflection.

The joint-writing process is an important feature of case reporting as a teaching method. The act of jointly creating a written report gives new meaning to the project, since the report includes personal impressions and additional information gained from in-company meetings, visits, interviews and presentations. This process of new-meaning creation has to be facilitated. To this end, the cases in this book were written by young researchers, mostly doctoral students from various universities, in collaboration with Siemens managers. The doctoral students acted as coaches of the process of joint sense-making. It is important to realize that these "case coaches", as we called them, do not act as teachers, but as "teasers". Whereas teachers typically instruct, and provide information, "teasers" tease out the accumulated knowledge and experience, and facilitate the reflective process necessary to elucidate the merits of this experience for the future. They push the participants to reflect on experiences, to create usable knowledge, to reflect on the value of knowledge, to think about where and how to share knowledge. All case writing participants are thus simultaneously teachers and learners.

My research has uncovered the following reasons for the particular success of this process:

1. All those who took part in a project are questioned not only about what happened, but also about how it happened, because creating a narrative means providing a logical account of the events.

2. The group of case writers comprises case coaches, who are company internals or outsiders, as well as project members of the case company or team. This adds an important dimension as the coaches may play the devil's advocate, questioning and challenging the inside view of the project, thereby "teasing out" valuable insights that would otherwise not surface.

3. Since the case coaches do not participate in the project, they are expected to research details and try to foster an understanding of how things work in the company. This obliges the insiders to carefully explain details that they would otherwise take for granted. The "outsiders", in turn, contribute by helping to conceptualize, to define and express lessons learned, by supporting the writing process, but without providing solutions and answers!

4. During the "teasing" process, case coaches may become aware of tacit assumptions, rules and prevalent behavioral codes that are not consciously perceived. They may force participants to question (hidden) rules, to reflect on behavioral codes and put events in a broader context. Differences that are not otherwise obvious are thus revealed. Discussion of these differences during the case-writing process may create a new awareness of certain rules, habits and behaviors in the case organization itself, which are usually hidden below the surface.

During the joint-writing process, managers describe the initial situation regarding the case, the problems they needed to solve, challenges they faced and then evaluate the results of the project. Thereafter the group discusses questions, such as how to assess the "facts" presented to them, which features of the case are especially noteworthy, and what they hope to convey to the readers. People from different backgrounds, i.e. managers from the case company, partners, coaches, consultants and relevant employees provide input. Case writing therefore not only integrates a wide variety of different viewpoints, but the collaborative writing process also provides a final report that differs quite radically from a study written by a single individual. Collective writing is sometimes a painful and a long process as it implies discussions, different viewpoints and experiences, comparisons and finding acceptance in formulation and relevance of the knowledge to be transferred. That is why learning occurs in the writing team itself, thinking about the narrative of a case and the teaching, and not only in groups where the propositions, the processes, applications, tools, lessons learned are to be taught.

Case studies or bulleted points?

The final report not only differs significantly from a study written by a single individual, but also from the bulleted lists that pervade corporate life. Why then, is joint case-study writing a better way of recording experience and learning from it, than other established tools, such as project reports or lists of bulleted points?

This question can be answered in terms of a distinction between the two parts of the report, namely the pure case study and a type of teaching note. The reports of the present book have all used this structure of case study and teaching note. The case study tells the story in a descriptive fashion; it is a narrative in which the writers relate the circumstances in which the project started, the difficulties endured, the changes introduced by the project, the questions it raised, the obstacles faced and the range of options. In rich narratives, teams could even build in their thoughts about dealing with uncertainty, fear, emotion, myth and mystery, power, etc. The teaching note, however, has an instructive rather than a narrative function. It contains joint reflections on the lessons learned, the key success factors, reports failures, if any, and offers reasons for them, describes the results, and offers conclusions.

In this book, the teaching note is reflected in the key propositions that are provided towards the end of each case study. These key propositions offer the insights provided by the individual cases in a condensed manner. It is important to note, however, that such key propositions develop their full potential only when viewed against the background from which they emanate, i.e. the case study itself. The case study method is therefore doubly effective. Firstly, the task of writing the report impels the authors to reflect on the content of the project. Secondly, the creation of the teaching notes requires them to contemplate the lessons learned (what points do I want to teach or illustrate by means of this case?). Perhaps most importantly, the teaching note prompts the writers to help readers to discover how the case can be decontextualized, the implicit knowledge applied in a new context, put to work and multiplied and how others can best learn what they need to know.

The educational concept behind the teaching note is the conviction that the ability to teach the case, i.e. to explain it to others step-by-step, is the best possible evidence of the teacher's thorough comprehension. Only by teaching (or writing teaching notes) does knowledge become conscious and explicit, and can thus be transferred. There is much more to learning than the passive acquisition of information: an effective teacher elaborates on the subject matter, reflects on its significance, considers links with other topics, and chooses an appropriate teaching method. Knowledge only forms an effective basis for action if the material has been well taught. Put differently, you have to understand first before you can instruct and transfer knowledge.

A factual project report without a teaching element cannot offer this unique combination of narrative, analysis and instruction. It is clear that bulleted lists in company reports cannot convey this unique combination of analysis, instruction, and narrative. The latter is a key element in case studies and perhaps the key advantage of case studies over company reports.

A tool for putting different types of knowledge to work

Our question here is how to put knowledge to work. Fundamental to this question is sensitivity to various types of knowledge. From my experience, case writing is a tool that is very sensitive to various types of knowledge. In this section, I explain firstly, why case writing is uniquely well qualified for making tacit knowledge explicit. I then proceed to demonstrate how case writing also accommodates, what I call, task related, conceptual and relational knowledge, thereby further contributing to putting knowledge to action.

Narrative case studies put tacit knowledge to work

A key element of case writing is that the product, i.e. the case report, is not simply a document containing data, information and knowledge, but can also be read as a story. Telling a story involves more than presenting facts in a logical order. In the case study, writers often use literary devices such as metaphors, vivid imagery, associations and characterization. The metaphor conveys a different and deeper layer of meaning that would be lost in a more factual report, simply noting the presence of "considerable differences between the two company cultures". These and other stylistic devices add meaning to the case study, offering the reader access to the project at a different level. If we read the case report as a narrative, and consider the images, metaphors and character descriptions, we find a layer of meaning containing knowledge that would otherwise remain hidden or tacit.

The distinction between tacit and explicit knowledge is indeed a critical one. Tacit knowledge is often bound to particular experiences and, furthermore, inherently bound to the person possessing it. It is traditionally assumed that transferring tacit knowledge implies transferring people. Unlike tacit knowledge, explicit knowledge can easily be transferred indirectly, i.e. via media such as telephone, electronic mail communication, and so on. Given the cost implications associated with the problem of transferring tacit knowledge, this can be seen as a problem of making tacit knowledge explicit.

My research shows that case writing is uniquely well suited to making tacit knowledge explicit. Within groups and organizations, shared meanings and tacit knowledge are defined in experience only, often not open to or part of the language in use. The labels, images and expressions used in the case studies are the outcomes of a process of constructing shared meanings. The report, due to its narrative style, conveys case writers' assumptions about how to describe aspects of a given project. The writers discuss ways of evaluating the results of the project, and of making sense of their own impressions, together with all the information gathered from documents, interviews and other sources. Members of the company use strategic narratives in their planning processes in order to clarify to others the thinking behind their plans, and also to capture the imagination and stimulate the enthusiasm of other employees. This technique is based on the recognition that a story defines a set of relationships and a sequence of events, and identifies causes and effects. The story weaves all these elements into a complex whole that is likely to be remembered.

Case writing reveals related, conceptual, and relational knowledge

Beyond the tacit or explicit distinction, another level of distinctions can be made between task-related knowledge, conceptual knowledge and relational knowledge. These three types of knowledge differ in scope, level of specificity, and degree of explicitness.

Task-related knowledge is factual knowledge that tells us how to accomplish a given task. This type of knowledge is highly specific. Its scope – i.e. its area of application – is fairly limited. It tends to be relatively implicit, because it largely consists of routine actions. An example would be an engineer's knowledge of how to build or repair a particular technical device.

Conceptual knowledge has a wider scope, but is less specific. It is concerned with ways of approaching a problem or a project. Good examples of conceptual knowledge would be typical procedures for launching a product, or for implementing a research and development project. In its broadest form, conceptual knowledge is knowledge about methods for solving problems. It is usually explicit, and provides a framework within which specific tasks can be approached.

Relational knowledge is mostly implicit, and relates to particular persons, habits, "the rules of the game" and hidden rules within an organization. Its scope is relatively limited, though some generalization is possible.

Again, my experience shows that the narrative style of case writing can reveal all three kinds of knowledge. Task-related knowledge is found in the description of a given project, the design of the constituent tasks and the challenges faced. Conceptual knowledge is conveyed in the description of the project procedure and its phases. Relational knowledge is the most difficult type of knowledge to record, but here the narrative aspect of the case study becomes useful. Elements of relational knowledge may be found, for example, in descriptions of the management interfaces between different divisions or geographically divided teams, or in the characteristics attributed to persons who played an important part in the project. The use of images, metaphors and associations may also convey relational knowledge. Discussion of the case also leads to development of the collective knowledge base, since participants contribute their personal insights, recall their own work experience, and add the impressions they have gained during visits to the company.

A tool for promoting organizational learning

Besides being a tool for creating, making explicit and transferring knowledge, and being sensitive to various types of knowledge, perhaps the most important benefit of case writing is the stimulation of organizational learning – the development and efficient use of the organizational knowledge base. It is efficient because the knowledge gained is used more often, the company is not reinventing the wheel, and practices are further developed and multiplied. In fact, through its sensitivity to various types of knowledge in the transfer process, case writing is likely to yield considerable learning effects at the individual, group and company levels.

Case writing is a useful tool for promoting learning at different levels in the organization. Some companies have already extensively used this approach. Holderbank created several best-practice cases, videotaped show cases and material for internal management development, some of which were later published by the Geneva Knowledge Group team consisting of Gilbert J. B. Probst, Steffen Raub and Kai Romhardt. British Petroleum installed a group for Post-Project Appraisal to collect information and write up case studies and lessons learned, as Frank Gulliver reports in the Harvard Business Review. Xerox published short cases on Knowledge Management tools, written by knowledge managers, specialists, consultants or project team members. Xerox also reviewed change programs and restructuring projects and created a workbook to stimulate further reflection, as David Garvin of Harvard Business School reports.

Learning at the individual level

The smallest unit of learning is the individual and this has implications for collective learning.

An important element in the learning process at the individual level, is the ability to make hidden organizational rules explicit and to reflect on them, thus revealing deep cognitive structures and making behavior patterns of the project teams or a company understood. Chris Argyris, a psychologist teaching at the MIT, has shown that case writing can be an effective tool for revealing the differences between espoused theories and theories-in-use, i.e. as they are actually practiced by individual members of the organization. In his method, participants are asked to write a two-page case conversation. A "case" in this context is an important situation experienced personally by the participants and relating to a problem that they have tried to solve within their organization. They are then asked to describe the steps taken to resolve the problem, including conversations with other members of the organization. One of these conversations is then chosen as the "case conversation".

In the right-hand column of a page used to describe the steps taken, the participant records the actual conversation. In the left-hand column, he or she notes the thoughts and feelings experienced during the conversation. A comparison of the two columns and a discussion of the difference between what is said and what is thought, or felt, reveal the discrepancies between theories-in-use and espoused theories.

In his role as moderator, Argyris explains how individual learning is inhibited by the gap between what is actually said and what is thought and felt. Making individual participants aware of this gap allows them to identify unexplored learning opportunities, both for themselves and for those with whom they discuss it. Argyris concludes that pre-thought and afterthought, i.e. reflection, can enhance learning on both the individual and group levels. Individual learning constitutes the logical foundation for learning at group level, but organizational learning is more than just the sum of individual lessons learned.

Learning at the group level

We shall consider learning processes in two groups: the group consisting of all involved in writing the case, and the project group members interviewed for the purposes of the case study. In the case-writing group, new knowledge is likely to be created through group discussion. The case coaches, especially, play an important role in stimulating discussion and ensuring that common ground does not emerge prematurely, in their role of devil's advocate. In this process, it is critical that individual views and perspectives are influenced and broadened by the group. This mutual challenging of viewpoints in discussions and the new knowledge that the process yields, make learning at group level more than just some aggregate of individual learning. The results of the process of joint reflection will probably differ from the thoughts of a single case writer. Joint thinking may be expected to yield additional insights for the members of the case-writing group. It is also interesting to note how far the actual writing of the case contributes to learning within the group. If "writing is thinking", then joint writing is joint thinking. This suggests that while individual learning is the starting point for group learning, group learning is more than the sum of individual learning, suggesting added benefit for the individuals.

My experience shows that participants will profit in two ways. First, the procedure gives each member of the group a chance to reflect on the project, and this is likely to reinforce individual learning. There is usually no other such opportunity to reflect on the meaning of past events, as members of the project team are immediately assigned to new groups and have new tasks to perform once a project ends.

At group level, however, the project group will receive feedback on the comments they made during the interviews, since the finished case report will be made available. They will then see how their points of view have been incorporated into the case study and what they added to the final picture. These project team members get the chance to review, to compare, and to discover their points in the larger context, the case as a whole. Each member of the project group will also be able to read other group members' complementary views. A similar process evolves at the company level.

Learning at company level

The question now is how does learning at individual and group level lead to organizational learning, or learning at company level? The problem is that, while there are many ways in which individuals and teams can learn from their project experience, there are, unfortunately, few mechanisms that companies can use to ensure that this learning is utilized at organizational level. Organizational-learning mechanisms are structural and procedural arrangements that allow organizations to collect, analyze, store, disseminate and use information in a systematic fashion. These mechanisms become embodied in organizational routines, practices and beliefs, and are considered important for improving organizational performance in a number of areas. Organizational learning can be seen as the sum of changes in the use of the organizational knowledge base.

The current compression of product and communication cycles, blurring of industry boundaries, and deconstruction of value chains, limit the time frames available to companies to gain revenue through their knowledge bases. We have seen that in the current economy, which is "blown to bits" by the new economics of information, that agile organizational learning could develop into the key competitive advantage for incumbent companies. Start-up companies often find it difficult to compete with large, established companies, because they lack the rich knowledge assets and experience these companies have developed in the past. And yet, precisely this knowledge base becomes obsolete in time horizons that approach zero asymptotically. The key question in this scenario is how to ensure that the organizational knowledge base actually evolves and is used at all times.

Again, my experience shows that the writing and use of cases can contribute to the evolution of the shared knowledge base. The process of feeding the written cases back into the company is an institutional arrangement for collecting, storing and disseminating information. When the case report is presented and distributed throughout the organization, it increases that organization's ability to reflect on its past, thus creating an environment conducive to collective learning. These days a company with a superior capacity for learning, possesses a key competitive advantage. The use of case studies can therefore ultimately enhance the competitiveness of incumbent companies relative to start-ups.

What does this mean for your company?

Let me finish where I started off – with the search for effective methods for managing organizational knowledge. It is common wisdom that the new economy calls for uniquely well-suited approaches for the management of organizational knowledge. The implications of the organizational knowledge of incumbents developing into the key strategic lever in the battle against start-up companies, are far-reaching, but seldom followed through in practice. IT has unfortunately not yet reached the level of sophistication necessary to accommodate all the types of knowledge (and may never reach it, as the specifics of knowledge will always include experience, which cannot be made explicit!) and their implications for Knowledge Management and organizational learning.

I suggest using case writing as a convenient alternative. Because case writing is narrative in style, it is more sensitive to the different types of knowledge than traditional knowledge management systems. Writing stories, describing situations and making reflections on actions, explicitly foster the retention of the knowledge transferred. Further, the devil's advocacy as a characteristic element of the case-writing process ensures that the knowledge transferred is subject to appropriate scrutiny. The scrutiny and joint sense-making that go along with case writing, further enables organizational learning on various levels, starting at the individual level, proceeding to group and company levels. The important point is that these three levels interact in the process, stimulating reflection on each level. Case writing concentrates on company specifics, problems, culture, strategy, etc. People involved in case writing think about teaching the lessons learned, the best practices and experiences, and create and share knowledge by reflecting and teaching in their Community (of practice). The outcome of this is enhanced organizational-learning capabilities, which are sorely needed in today's new economy.

Writing cases after the action, in the sense of post-project-appraisals, after-action reviews, lessons learned, means that Knowledge Management and organizational learning become part of a continuous improvement process. In order for case writing to be fruitful for a company, several success factors need to be considered.

1. Collaboration between company "insiders" and "outsiders" is important, because it is through the interplay of questions and discussion within the case-writing group that knowledge becomes conscious. The "outsider" especially should practice being the devil's advocate extensively.

2. A narrative style should not only be permitted, but also actively encouraged. Managers who are used to writing reports or executive summaries in an objective style may need encouragement and training to adopt this style. Without the narrative element, the benefits of case writing are seriously compromised.

3. Case writing requires collaboration not only amongst individuals, but also at company level. Companies must be prepared to disclose and openly discuss the challenges and experiences of the past if they are to learn from them. The case-writing method can only achieve its full potential in companies where this tolerance is present, and where there is a genuine desire to profit from experience.

4. Case writing needs sponsorship and visibility, but also the willingness of top managers to use the lessons learned, the teaching, and the knowledge that is in the cases.

References

Argyris, C. (1982). Reasoning, learning and action. San Francisco: Jossey-Bass.

Argyris, C. (1997). Learning and teaching: A theory of action perspective. Journal of management education, 21 (1): 9-26.

Beer, M. and Eisenstat, R. (2000). Overcoming the "silent killers" to strategy implementation and organizational learning. Harvard Business School working paper, forthcoming in Sloan Management Review, Summer 2000.

Evans, P. B. and Wurster, T. S. (1997). Strategy and the new economics of information. Harvard Business Review, September-October: 71-82.

Garvin, David A. (2000). Learning in Action, Boston: Harvard Business School Press.

Garvin, D. A. (1993). Building a learning organization. Harvard Business Review, 4: 78-91.

Gulliver, Frank R. (1987). Post-Project Appraisals Pay. Harvard Business Review, 65: 128-132.

Probst, G. J. B.; Raub, S. and Romhardt, K. (2000). Managing Knowledge, London: Wiley, first published as: Wissen managen. Frankfurt: Gabler, 1997.

Probst, G. J. B. (1998). Practical knowledge management: A model that works. Prism, 2: 17-29.

Written by

Notes on the contributors

Christina Bader-Kowalski

completed her apprenticeship with Siemens Switzerland. In more than 20 years she has broadened her knowledge and experience in different departments such as Sales, Marketing and Human Resources. In the latter her fields of functions are located in Languages Trainings, e-Learning and Management Development. Currently she is responsible for e-Learning within Siemens. This includes the implementation of e-Learning strategies as well as the development of e-Learning modules and tools for the intranet.
She can be contacted at: christina.bader-kowalski@cp.siemens.de

Dagmar Birk

joined Siemens in 1995 and worked in various functions such as software engineering and product management. She is now responsible for the marketing of the Knowledge Sharing group of Siemens Medical Solutions.
She can be contacted at: dagmar.birk@med.siemens.de

Thomas H. Davenport

is Director of the Accenture Institute for Strategic Change, and Distinguished Scholar in Residence at Babson College. He writes and speaks on the topics of information and knowledge management, reengineering, enterprise systems, and the use of information technology in business. He has a Ph.D. from Harvard University in organizational behavior and has taught at the Harvard Business School, the University of Chicago, and the University of Texas. He has also directed research at Ernst & Young, McKinsey & Company, and CSC Index. He has recently co-authored *The Attention Economy* (Harvard Business School Press), which describes how individuals and organizations can manage "the new currency of business." Prior to this, Tom wrote, co-authored or edited eight other books, including the first books on business process reengineering, knowledge management, and enterprise systems. He has written over 100 articles for such publications as *Harvard Business Review, Sloan Management Review, California Management Review,* the *Financial Times,* and many other publications.
He can be contacted at: thomas.h.davenport@accenture.com

Marc D'Oosterlinck

studied electromechanical engineering at the Hogeschool Ghent, Belgium. He joined Siemens NV Belgium in 1981 and worked in various functions, the previous being

Engineering Manager. Currently he is Know-how Manager at Siemens AG, Erlangen, Industrial Services Division Middle and East Europe I&S IS.

He can be contacted at: marc.doosterlinck@siemens.com

Andrea Dora

gained four years' extensive business know-how of telecommunication solutions in the wholesale and retail client sections as sales manager at Siemens in the eastern region after her final examinations and business science study. During this time she was involved in the task force "debt collecting" at Information and Communication Networks, Sales Germany. As knowledge manager she has for the past 3 years lead initiatives for the implementation of Knowledge Networking at ICN VD in München. For example, KN Service Knowledge – an established process for knowledge networking of all employees in the service environment – has received special attention across business boundaries.

She can be contacted at: andrea.dora@icn.siemens.de

Ellen Enkel

is project leader at the research center KnowledgeSource (www.knowledgesource.org); a cooperation of the Institute of Management (Prof. Georg von Krogh) and the Institute of Information Management (Prof. Andrea Back) of the University of St. Gallen. The aim of KnowledgeSource and especially the Competence Center Knowledge Networks for Growth that she leads is to develop instruments for knowledge management in cooperation with researchers and companies. She is especially involved in developing a method of establishing and maintaining such networks (MERLIN method) with a focus on fostering innovation and strategic management aspects.

She can be contacted at: ellen.enkel@unisg.ch

Christine Erlach

psychologist and consultant for integrative knowledge management, worked at the Institute of Educational Psychology at the University of Munich. She was one of the didactical designers of the Knowledge Master; at the pilot and during two trial runs she was responsible for conducting the presence workshops and tutoring the virtual teams. She is currently conducting a research- and consulting-project about narrative knowledge management and is writing her doctoral thesis about Story Telling as a KM-tool for cultural change.

She can be contacted at: erlach@edupsy.uni-muenchen.de

Michael Franz

is a doctoral candidate at the University of Graz and the University of Mannheim. He worked for a number of companies including Siemens, Deutsche Bank and KPMG. He was awarded his Master's degree in Commerce (Dipl.Kfm) at the University of Mannheim. He lives in Munich and is primarily engaged in coaching and training projects.

He can be contacted at: michael_franz@gmx.de

Hartmut Freitag,

studied computer science at the University of Kaiserslautern and received a Ph.D. from the University of Hamburg. After 10 years of research activities at Siemens Corporate Technology, he is currently responsible for technology and innovation management at Siemens Industrial Services in Erlangen.

He can be contacted at: hartmut.freitag@siemens.com

Kurt Freudenthaler

studied mathematics at the University of Erlangen-Nuremberg. For more than 20 years he worked in different functions (software development, IT strategy, logistics) at the Siemens Power Generation Group. From 1997 until 1999 he was project leader of the Siemens Westinghouse IT integration team. In 2000 he joined the Corporate Knowledge Management office and was at first engaged in setting up the Knowledge Community Support. Now he is responsible for the Knowledge Management rollout.

He can be contacted at: kurt.freudenthaler@siemens.com

Tanja Gartner

studied communications, business administration and sociology at the University of Zurich in Switzerland. Her final thesis focused on Knowledge Management in Strategic Alliances with a case study of the Star Alliance. She is now working as a consultant and member of the Knowledge Management Practice at Siemens Business Services. She was responsible for piloting Siemens Business Services' new knowledge management approach in Germany.

She can be contacted at: tanja.gartner@mch20.sbs.de

Michael Gibbert

is an in-house consultant in knowledge management and strategy at Siemens AG, Munich, and a doctoral candidate at the University of St. Gallen. In 2002, he is on study-leave at the Yale School of Management. Prior to joining Siemens, he worked as a research Associate at the European Institute for Management Studies (INSEAD), and at the University of Stellenbosch, where he was awarded his Master's degree in Commerce. He lives in Carmine Superiore, Italy and wonders why nobody else does.

He can be contacted at: michael.gibbert@icn.siemens.de

Albert Goller

is president of the Center of E-Excellence at Siemens. The Center of E-Excellence is engaged in initiating new processes in e-business, and helping both business groups and regions to evolve to e-business. His background is in engineering in which he earned his Masters Degree (Dipl. Ing.) with a major in communication/data. He has 27 years of business experience in the following areas: sales and marketing, commercial/residential construction business, communications and information, and e-business. Prior to joining the Center of E-Excellence, he was President and Chief Executive Officer Siemens in Canada, and vice President of I-Center, Siemens Erlangen, the distributor for Germany, Austria, Poland, Hungary, the Czech Republic, and Slovakia. He serves on various corporate directorships and associations, including the Canadian/German Chamber of Commerce of which he is Vice Chairman.

Joachim Graff,

studied electrical engineering and economics at the Technical Universities of Giessen and Munich. Main activities: Internet and Intranet marketing, and operational tasks for knowledge management tools. He is the Know-how Manager in the area of General Contracting at Siemens AG Erlangen.

He can be contacted at: joachim.graff@siemens.com

Irmgard Hausmann

is Project Manager for Knowledge Master at Siemens Qualification and Training (SQT). She holds an Engineering degree and has many years' international experience in further education of Information and Communication Technology. She was responsible for the conception, development and marketing of innovative training technologies and headed various European research projects on distance learning. Currently she consults on E-Learning projects, especially on Knowledge Management.

She can be contacted at: irmgard.hausmann@sqt.siemens.de

Peter Heinold

started his career at Siemens as an industrial apprentice in 1978 and since then has worked for Siemens in various positions, functions and locations. He held positions in the fields of computer sales, in corporate communication and PR in Canada, and worked as controller, strategic planner and international sales manager for the Defence Electronics Group and as a business analyst in the Corporate Planning and Development department (CD). Since 1998 he had also represented CD as a member of the Corporate Knowledge Management (CKM) taskforce. He is the founder of the CKM council and co-founder of the CKM office. Since the start of the CKM office in October 1999, he has worked as an internal and external consultant for Knowledge Management and as program manager for the Marketing and Promotion of CKM, as well as for the implementation and development of the CKM bodies. He is a member of several international KM communities such as the KNOW Network, KMCI, GKEC and the German "Industriearbeitskreis Wissensmanagement in der Praxis", (www.wimip.de). He is invited speaker at many KM conferences and meetings.

He can be contacted at: peter.heinold@siemens.com

Dr. Josef Hofer-Alfeis

studied communications and received his doctorate degree in the fields of pattern recognition and image development from the University for Applied Sciences in Munich, Germany. He joined Siemens AG in Munich in 1984 and worked on innovative industrial applications of document management, knowledge-based systems and business process models. He has also been researching issues regarding systematic knowledge management since 1990. Dr. Hofer-Alfeis has acted as Moderator and Spokesperson for the Siemens-wide KM Community of Practice, with over 450 members worldwide. He is a member in the Siemens Knowledge Management Corporate Center, which was set up by the Siemens AG Central Executive Committee upon the Community's recommendation and which is responsible for the following programs at the company level: strategy and metrics as well as knowledge management and competence development.

He can be contacted at: josef.hofer-alfeis@siemens.com

Antonie Jakubetzki

is the regional Learning Manager Europe, Africa and Middle East for the strategic, worldwide Siemens Management Learning Programs (S1-S5) and Program Director S3, which is carried out in partnership with Duke University and Babson College. Before joining Corporate Human Resources in 1997, she was active in a number of capacities for Siemens AG: in international sales, head office for strategic functions and, lastly, in the development of the Siemens top initiative worldwide. She was actively involved in the overall design process of Siemens Management Learning and is also in charge of Siemens' internal marketing of the programs.

She can be contacted at: antonie.jakubetzki@cp.siemens.de

Claudia Jonczyk

is currently a research assistant and Ph.D. candidate at the University of Geneva after having completed her Master's degree at the University of St. Gallen. At present she is researching organizational learning and the use of organizational narratives at the MIT Sloan School in Boston, MA. She has worked extensively on several knowledge management initiatives at Siemens ICN.

She can be contacted at: jonczyk@mit.edu

Susanne Kalpers

is a senior consultant at Siemens Corporate Technology, Knowledge Management, and specializes in IT aspects of mergers and acquisitions. She has been involved with the MAKE project from the onset.

She can be reached at: susanne.kalpers@mch.siemens.de

Marla Kameny

is currently a Research Associate and Ph.D. candidate at the Institute of Management at the University of St. Gallen, Switzerland, after having completed her MBA and Masters in Industrial and Labor Relations at Cornell University, Ithaca, New York, USA. Her professional experience is in the area of Human Resource Management within various industries. Her current research addresses the field of strategic management and capability building, with a focus on growth and organization development in high-growth, start-up environments.

She can be contacted at: marla.kameny@unisg.ch

Klaus Kastin

has been working for the Strategy and Business Alliances department at Siemens Information and Communication Mobile since 1998, with a specific focus on strategy and corporate reporting.

He can be contacted at: klaus.kastin@mch.siemens.de

Birgit Kleiber

has worked in different functions at Siemens Medical Solutions since 1988. She has been working at the Center of E-Excellence since October 2000 and is currently responsible for communications in the e-business project "e-Community Knowledge Management", which is jointly run by experts from Corporate Knowledge Management (CKM) and Corporate Technology, Information & Communication (CT IC). She is a member of different global Knowledge Communities on e-business related topics within Siemens and completed the Knowledge Master certificate in July 2001.

She can be contacted at: birgit.kleiber@e.siemens.com

Thomas Klingspor

is currently a student of business engineering at the Karlsruhe University. He helped to conceptualize and implement the pilot of the Knowledge Networking Enabling process during a 6-months internship with Siemens' ICN VD knowledge networking team. Currently he is working as a student assistant at the Institute for Organizational Studies and Strategic Management (IBU).

He can be contacted at: thomas.klingspor@gmx.net

Felix Klostermeier

is currently a knowledge networker at the Siemens Information and Communication business unit. After studying Business Education, he started working with Siemens as a sales manager, but found his niche at Knowledge Networking. He redesigned and rolled out the Yellow Pages that produced the Competitor Analyses and now works as an in-house knowledge consultant for the knowledge networking team. He focuses on employee-centred, but still measurable, knowledge management.

He can be contacted at: felix.klostermeier@icn.siemens.de

Hartmut Krause

has worked for Siemens since 1973 in different functions at the Medical Engineering Group Med (development, marketing and innovations and strategy process). In 1999 he was the project leader at Med of the project "Best-Practice Marketplace". Since the end of 1999 he has been program manager at Corporate Knowledge Management (CKM) where he is responsible for Best-Practice Sharing.

He can be contacted at: hartmut.krause@siemens.com

Petra Kugler

Is a Ph.D. candidate and Research Fellow at the Institute of Management (IfB) at the University of St. Gallen. Currently, she is also a Visiting Scholar at the University of California at Berkeley. She holds a Bachelor and Diploma Degree which she obtained at Augsburg University, Germany and during an exchange year at the Università degli Studi di Bari, Italy. She has worked in the marketing and advertising areas. Her current research addresses the field of strategic management and organization, mainly focussing on new forms of organizing such as open source software development, knowledge and capability building, and innovation.

She can be contacted at: petra.kugler@unisg.ch

Marius Leibold

is Professor in Strategic International Management at the University of Stellenbosch, South Africa, and the Business School Netherlands. His research focuses on new business models for global competitiveness, and complexity management approaches for corporate innovation and organizational reinvention. He advises corporations and industries on global competitiveness, especially on measuring organizational fitness for global competitiveness in fast-changing industry landscapes. The industries he has advised regarding global competitiveness of countries and regions include the wine, airline, insurance, tourism and life sciences industries. He collaborates with Harvard Business School's Center for Organizational Fitness and various other prominent international research and consulting organizations.

He can be contacted at: leibold@mweb.co.za

Heinz Mandl

is Professor of Education and Educational Psychology at the Ludwig Maximilian University of Munich. His broad scientific research is based on modern constructivistic approaches to teaching and learning environments such as situated cognition. Within this theoretical background he also is engaged in the design and evaluation of virtual learning environments and the further education of Knowledge Management in organizations.

He can be contacted at: mandl@edupsy.uni-muenchen.de

Manuela Mueller

studied in Business Administration at University of Applied Sciences and Research in Munich. Post Graduate Studies in General Management and International Marketing at Università degli Studi di Roma "La Sapienza", Rome, Italy. Consulting activities in Italy. Manager in the Knowledge Management team at Siemens Industrial Products and Technical Services, Erlangen. Manager Knowledge Management at Siemens Health Services GmbH & Co. KG, Erlangen. She lectures on Knowledge Management at the University of Applied Sciences and Research in Munich. She is the newly appointed Head of Knowledge Management, Siemens Medical Engineering Group.

She can be contacted at: manuela.mueller@siemens.com

Hans Obermeier

has a Ph.D. in human biology from the Ludwig Maximillian University in Munich and worked as a scientist for five years. Thereafter he became a consultant for marketing and communication, focusing on Knowledge Management, at an agency in Munich. His current position is Management Consultant at Siemens Business Services. As a member of the knowledgemotion™ project team he is responsible for the deployment and internal marketing of the system at SBS.

He can be contacted at: hans.obermeier@mch20.sbs.de

Karoline Petrikat

has been working as consultant for the Strategy and Business Alliances department at Siemens Information and Communication Mobile since 1998. As project manager she was involved in several M&A transactions. She was responsible for the Mergers & Acquisition handbook and has been part of the MAKE project from the onset.

Gilbert J. B. Probst

is professor of organizational behavior and management and director of the Executive MBA program at the University of Geneva, Switzerland. He received his Ph.D. as well as his habilitation in business administration from the University of St. Gallen, Switzerland where he was also a lecturer and vice president of the Institute of Management. As a visiting faculty member he taught at the Wharton School of the University of Pennsylvania, Philadelphia, as well as at the International Management Institute (IMI) in Geneva. Gilbert J. B. Probst is the founder of the Forum of Organizational Learning and Knowledge Management with 16 top companies and founder and partner of the Geneva Knowledge Group, a consulting company. He is a consultant for various major companies. Prof. Probst is a member of the board of Kuoni Travel, an important European tourism group, HOLCIM, one of the world largest cement producers, and Alu Menziken Holding, Aluminium extrusions – in US, Asia and EC. His publications deal with the subjects "Learning Organizations", "Knowledge Management", "Joint Venture Management", "Thinking in Networks" and "Structuring of Organizations".

He can be contacted at: gilbert.probst@hec.unige.ch

Nicole Prummer

is currently Dealflow Manager of the Siemens ICN nCubator, supporting venture teams with high potential business ideas. During her time at Siemens ICN, she was actively involved in rolling out ICN ShareNet internationally. Her special focus was on implementing knowledge related metrics as well as incentives and structures for international collaboration. As a project manager she set up the strategic controlling system for the ICN board, the ICN Balanced Scorecard.

She can be contacted at: nicole.prummer@icn.siemens.de

Dirk Ramhorst

studied Business Administration, specializing in Innovation Management, at the Christian Albrecht University in Kiel. He has been a management consultant for IT strategy at Preussag. His current position is Chief Knowledge Officer at Siemens Business Services. As the CKO and leader of the knowledgemotion™ project team he is responsible for the implementation of an SBS-wide Knowledge Management system.

He can be contacted at: dirk.ramhorst@hbg.siemens.de

If you would like to have a copy of the framework diagram (poster size – DIN A2), please feel free to contact us via info@knowledgemotion.sbs.de

Rainer Schmidt (†)

Rainer Schmidt was a consultant at the knowledge management department of the Corporate Technology Division of Siemens AG. He was actively engaged in the KECnetworking project. His research focus lay in the area of developing and supporting product communities. He died tragically in August 2000.

Stefan Schoen

has been working at Siemens Corporate Technology since 1996. Until October 2001 he was doing consulting for the implementation of knowledge management solutions at different Siemens business groups and the Siemens Knowledge Community Support. Since Nov 2001 he is head of the department "User Interface Design" with locations in Munich, Princeton and Beijing. His Ph.D. thesis focussed on the design and support of Communities of Practice. He has engineering and business administration degrees from the Technical University of Munich, the University of Stuttgart and the University of Wisconsin, USA.

He can be contacted at: stefan.schoen@mchp.siemens.de

Sabine Seufert

is senior lecturer and project manager at the Institute for Media and Communications Management (MCM Institute) at the University of St. Gallen. She is responsible for a new MBA program Executive MBA in New Media and Communication. Her research focus lies in the area of new media in education and knowledge management.

She can be contacted at: sabine.seufert@unisg.ch

Jörg Späth

has worked in the field of Internet and Intranet solutions since 1995. Currently he works for Siemens ICM's e-business department.

He can be contacted at: joerg.spaeth@mch.siemens.de

Rob van der Spek

is principal consultant knowledge management at CIBIT Consultants|Educators located in Utrecht, the Netherlands. CIBIT is an independent European consulting and